Immortal

Immortal
Our Cells, DNA, and Bodies

David Goldman

ELSEVIER

ACADEMIC PRESS
An imprint of Elsevier

Academic Press is an imprint of Elsevier
125 London Wall, London EC2Y 5AS, United Kingdom
525 B Street, Suite 1650, San Diego, CA 92101, United States
50 Hampshire Street, 5th Floor, Cambridge, MA 02139, United States
The Boulevard, Langford Lane, Kidlington, Oxford OX5 1GB, United Kingdom

Notices

Knowledge and best practice in this field are constantly changing. As new research and experience broaden our understanding, changes in research methods, professional practices, or medical treatment may become necessary.

Practitioners and researchers must always rely on their own experience and knowledge in evaluating and using any information, methods, compounds, or experiments described herein. In using such information or methods they should be mindful of their own safety and the safety of others, including parties for whom they have a professional responsibility.

To the fullest extent of the law, neither the Publisher nor the authors, contributors, or editors, assume any liability for any injury and/or damage to persons or property as a matter of products liability, negligence or otherwise, or from any use or operation of any methods, products, instructions, or ideas contained in the material herein.

Library of Congress Cataloging-in-Publication Data
A catalog record for this book is available from the Library of Congress

British Library Cataloguing-in-Publication Data
A catalogue record for this book is available from the British Library

ISBN 978-0-323-85692-8

For information on all Academic Press publications
visit our website at https://www.elsevier.com/books-and-journals

Publisher: Andre Gerhard Wolff
Acquisitions Editor: Peter B. Linsley
Editorial Project Manager: Barbara Makinster
Production Project Manager: Punithavathy Govindaradjane
Cover Designer: Mark Rogers

Typeset by SPi Global, India

Working together
to grow libraries in
developing countries

www.elsevier.com • www.bookaid.org

To Barbara and Armond Goldman

Contents

About the author

Born in Galveston in 1952 into a medical family, David Goldman was acutely aware of mortality from about the age of five. He and his brothers and sister were not afraid of the dead, but maybe death was a different matter—they held their breaths every time they passed the big cemetery on Broadway. Any such fears never stopped them living life on its edges—whether swinging from tree to tree like apes or surfing hurricane waves. David Goldman graduated with honors from Yale and the University of Texas Medical Branch in his hometown. As a medical student he was an extern at the National Institute of Mental Health, and after a brief residency returned to NIMH as a postdoctoral fellow and clinical associate before establishing an NIH lab in 1986. Presently, he is Clinical Director and Chief of the Lab of Neurogenetics of the National Institute on Alcohol Abuse and Alcoholism. He is one of the most highly cited scientists in neuropsychiatric genetics, having published studies connecting genes to behavior in leading journals such as *Science* and *Nature*. His academic honors include the top award of the Research Society on Alcoholism and he is a Fellow of the American College of Neuropsychopharmacology. He chaired an NIH human research protocol review board (IRB), and the Scientific Review Committee of two NIH institutes. He taught human genetics as an Adjunct Professor at George Washington University. For many years he has lived in a unique house, designed and partly built by him, in the woods of Potomac, Maryland where he bicycle commutes, cleans roads, herds cats, and helps a forest sequester carbon in the rhythm of its seasons. He has three children and three grandchildren. His wife of 15 years, Nadia Hejazi, is a pediatric neurologist.

Introduction

A live body and a dead body contain the same number of particles. Structurally,
there's no discernible difference. Life and death are unquantifiable abstracts.
Why should I be concerned?
Alan Moore, attributed to Dr. Manhattan in Watchmen

Near her last day, my mother asked me, "How do I get out of this?" Memory burdened her more than all her maladies of age. Her ghosts outnumbered the living. She was 11th of 12 children, the others gone. Where were her Texas longhorns, Windy Hill farm, or little dog? That dog once nearly killed itself by leaping from a balcony after a bird. After she was old and her arthritis stopped her from living like she wanted, my mother liked to stand for a long time looking out to sea from her high balcony. In the witch's fortress somewhere far above Kansas, the little dog escapes and someone inverts the red hourglass. Ruby slippers magically return the girl, and the shoeless dog, to a high plains homeplace of endless fields and heaven above. Mom was blameless if she came to outlove the living for the dead. In our later years, we begin looking back in time more than to the future or the present. For example, Bobby was born years before the rest of us kids, but year by year that we lived after him he became another year younger than us. Every year Mom lived after her kid sister Billie was a bittersweet trick of memory.

This book is partly about memories of the dead that are both joyful and sorrowful. In places it is a bit recursive and self-referential, perhaps as a reminder of how memory works. The void created by loss of loved ones is scarcely reduced by time, and the nth time memory is reawakened, emotions are rekindled along with a network of memories. Mom had learned to cope, and maybe this was why she was partly good at teaching us kids to reframe memories, soldier on, and thank life for irony and diversion. And, at last, to find a way out. A few days after she asked, her sorrow passed, easy as Texas twilight.

What is life? Conception or birth, as implied by the ways words are used every day, are sometimes reckoned the beginning. But human life is usually said to begin somewhere between. If a person begins at conception, should an embryo be treated with dignity? If so, where is the line drawn: newborn, viable fetus, fetus with a heartbeat, embryonic cell mass, pluripotent cell, cell that can be transformed into a pluripotent cell, or the genetic program? These are questions I ask as a mortal person, and as a physician scientist who studies DNA, cells, and the brain. Do the dead and unborn have a moral standing and does a brain, a memory, or a DNA molecule from which a person may be constructed? Why do we honor the dead, or the living?

Scientists propose (seriously and recently, and in the august journal *Science*) that inanimate rivers and mountains should be accorded legal rights on a par with people. The end-goal is good—to protect rivers and mountains. But what of the means? Life gives everyone willing to observe a knowledge of life and death that sets the boundaries of these things we call life and human life. When should a body be declared dead and taken off "life support"? The end-of-life tragedies of Nancy Cruzan, Terri Schiavo, Karen Ann Quinlan, and Jahi McMath illuminate dilemmas of modern medicine. Families and institutions take turns exchanging opposing sides as to whether the body is a person—a loved one—or the animated remains of a person. The body is immortal, but selves do not survive. In these end-of-life dramas some find meaning and solace, but a great many suffer and lose faith in institutions, each other, and faith itself. If everything is important, then nothing is important, and a definition of life encompassing rocks means that no lives matter. We should not tell people dealing with death that dirt thrown on a coffin is equivalent to what is inside. But we still must decide what are humans and what are things. A body in coma or cryostasis is not animate. A cell or embryo that can make a person is inanimate, and can also be kept in stasis or discarded as if of no more significance than a rock. This book attempts to add to the vast knowledge that anyone who has experienced life has about life, through the facts of life that science can teach, knowing in advance that future generations will know better. But all the facts of science that can be mustered only augment personal experience.

Bobby, who liked to be called Robert, was one of my blood relatives who died of bladder cancer. In a fit of undoing I underwent cystoscopy, a procedure wherein a urologist sticks a probe up into the bladder to see if anything is amiss—as if I had not suffered enough from kidney stones and infections that brought me close enough to death. For that matter, as "survivors" often ask, why had not I drowned as a 5-year-old in an obscure arm of Galveston Bay called Offatts Bayou? The obituary could have been enlivened by the amusing fact that the boy drowned because an inflatable shark eluded his grasp. Such are the ignominies of death. In truth, anyone living to old/not-so-old age is lucky not to have died in some strange or ordinary way. We dodge death from bad cars and bad drivers. Millions of cancer cells lurk in our bodies. Not to mention, and then to mention, details of personal experience: falling from trees, cramping while swimming far beyond the breakers, canoeing in hurricanes, bike crashes, being thrown from horses, inflatable sharks. Then too, as embryos we successfully implanted in a uterus, and our mothers did not abort us deliberately or spontaneously. Anyone living won the lottery by dodging a myriad spontaneous genetic mutations and teratogenic exposures that would have killed us, even if we were all born with disabilities. If our parents transmitted us favorable genotypes and took care of us, it was not because we chose them wisely. In my family, kids lived dangerously, and surprisingly only the youngest has died so far. If it consoles us, we can also claim Bobby was lucky to live into his 50s, and not confuse longevity with value. And with what book did I choose to comfort him? Why, Melville's *Moby Dick*, because, strangely, Bobby had only seen the movie starring Gregory Peck chasing a whale, or maybe unconsciously because I was almost killed by an inflatable shark.

Medicine teaches that experience is personal. Physicians categorize but know that each patient is individual. Intellectual understanding can be derived from textbooks, but empathic understanding is augmented by parallel experience, which is a knowledge we feel in our bones. The older we get, the more beloved friends we lose, and they are all irreplaceable and not merely representatives of a type. Experience teaches not to dwell on death, but the message of this book is to keep their memories as a legacy, the better to empathize with others. For me, the carnage includes many mentors and colleagues: Markku Linnoila (once a child movie star in Finland, a motorcyclist, classical guitarist, greatest psychopharmacologist on the planet, martinet), Anne Kelley, T.K. Li (a father of alcoholism research), Matt Reilly (first, a colleague by that name, and then—as I was finishing this book—my own son-in-law). Scientists and good friends in my lab, Michele Filling-Katz, a pediatric geneticist shot dead by her stepson; Richard Lister, gifted in intellect but unlucky in love. They all deserve biographies, but probably most will never be immortalized in Wikipedia.

It is not the unexpected tragedy that leaves the greatest impression, but death's approach. King Lear feared decrepitude and decay, not oblivion. The tragedy is not that death comes, but that it comes as a juggernaut whose advance we perceive too clearly. In this way humans are inferior to cats. Maybe cats know god, and they fear things, yes. But they do not fear death as a concept and in real life cats do not believe in the Heaviside Layer as a layer in the atmosphere where some may dwell after life, as in Andrew Lloyd Webber's musical *Cats*. They can hear the music but they do not know what Holst's "Neptune" means. They are like Adam and Eve, who occasionally suffered discomfort, what with living without clothes, shelter, or tools, but never feared death before they tasted the fruit of the tree of knowledge. To illustrate the benefits of freedom from awareness of death, I offer a Finnish parable and the story of a little bird.

According to Jaakko Lappalainen, once in my lab:

A young man is to marry tomorrow. He and his bride will share life and perhaps one day die among family and cherished memories. However, as Carly Simon sang and as the bachelor thought, usually it doesn't work out that way, and actually he would probably never be happier than he was at that moment. The night before, he therefore gets bombed at his bachelor party (polttarit). All his dreams and memories are happy. As a prank, his friends roll him up in a blanket and leave him on his fiancée's doorstep. The night is cold. The next morning she finds him, dead.

Is this story true, or a parable meant to be pathetic, ironic, or even humorous? David Lynch's protagonist in *Lost Highway*, and my dad, tended to remember facts in their own way. We all know such people. Maybe their memories are so good they remember things that never happened. My mom and myself are the type of people who have a different problem. We are scrupulous with facts but are confounded by the interpretations, even of little stories such as this.

So how about the little bird? It is a veery—a thrush, as is also the North American robin. Unlike robins, veeries are well-camouflaged, singing their hearts out unseen in deep woods. Every spring they return and at daybreak flit about, hyperaroused. The shadows are constantly changing and the terrain is unfamiliar, but they are as happy as they are foolish. It is precisely then that a veery smashes into a window, killing itself instantly. This happens every spring. Maybe their greater ability to avoid crashing into houses is one reason robins vastly outnumber veeries, but probably it is as simple as the fact that a robin is a marvelously preadapted suburban lawnbird. To one way of thinking, a veery's life is a happy story, but anyway not mainly sad, even if it comes into this world with sealed orders to crash into a window. Every year there will be young veeries and men and women with their whole lives in front of them, and at least partly oblivious.

Physicians see many patients near the end—the waters of life force ebbing. We worry if they are happy. When Bobby was a boy, and while I was a medical student in Galveston, Texas, my father-in-law died from ravages of brain cancer. His name was Kenneth Gemmell. I wonder whatever happened to his vast Masonic library. Perhaps his books illuminated mysteries, but I never read one. I got his stamp collection, a daughter of Scotland and Chile, and memories of the man. When I went down to Texas to be with Bobby, it was not the appearance of his body that destroyed me. Formerly so vital, he resembled a survivor of Auschwitz. Worse was a fatal resignation in his eyes that I had seen before, both in life and in art. As a young man, Jacopo Tintoretto was known as *Il Furioso*, and his self-portrait shows a man taking on the world. His self-portrait 50 years later captures the eyes of a man who had seen his destiny and the limits of his future (https://i2.wp.com/www. nationalreview.com/wp-content/uploads/2019/04/tintoretto-self-portraits.jpg?resize= 1024%2C636&ssl=1). My brother Bobby was a much lesser artist (and who is not?) and left less to posterity, but after all his life and all his personal journeys, he also knew well where he had been, what he had given, and where he was going.

I am grandson of a namesake, brother to four, twice married, father of three, grandfather of three, and was present at the death or memorial of many. I delivered babies and saw abortions. This is not special knowledge, but it does not need to be, to be motivation. Death and birth shatter complacency, whether it is the death of one or many, and for anyone sensitive the birth and death, and killing, of other life is also deeply touching. Many people, most who do not have posttraumatic stress disorder or any identifiable psychiatric disease, are profoundly moved. The dead visit in dreams.

It is hard to say that the birth and death of other animals matters less. I sacrificed animals for the sake of science but held fragile birds of all kinds to keep them safe from cats or cold. It seems inadequate to say that at least I do not eat other animals, when billions of vegetarians have found they can live perfectly well without animal flesh. Molecular biology proves that none of these creatures are "lower" on the evolutionary tree. Furthermore, in flight and grace they seem far more elegant and perfected than people, and it is hard to think that some are unaware of that fact. Oh, what a piece of work is a bird, and yet we eat them.

Should we punish people who kill by executing them? As a young man I would have said "sometimes," but as an old man, "never." Life is too precious. Societies, families, and individual psyches are torn by the status of fetuses, millions of which spontaneously abort and millions more that are deliberately aborted. Any woman making a choice to terminate her pregnancy does so within a social context. But what of the cells that could be embryos, and then fetuses, and people? If they too have status, we have a crisis made of multitudes of cells, including those of people such as Henrietta Lacks whose cells, known as HeLa, are cultivated in laboratories worldwide.

The only "settled science" about the boundaries of life is that there is no rational way to define them. Perhaps more than any other debate, the related questions of what is life and who is a person generate the strongest passions and demand the greatest sensitivity and respect. The conversation can at least encourage respect. Anyone coming to a book about mortality has emotions and opinions formed from a trove of experience. The purpose of this book is to add to anyone's vast knowledge based on what the sciences can teach. Causal chains tied to a person's physical being exist before conception and persist after. But with or without new technologies such as cloning, cryogenics, or cybernetics, does personhood continue after death of the body? Perhaps, even if a person has found oblivion rather than sanctuary. Spiritual, quasireligious, and religious experiences are places sciences cannot go and should not attempt to obliterate or replace. We do not need to discount the mysteries of reincarnation or afterlife, and unverified reports of the delirious at death's edge, to understand how all people were partly formed before emergence from the womb and continue to influence the world after being encrypted or turned to ash.

Genealogy enables intergenerational and transgenerational connections that are partly real and mainly illusory. Distant degrees of genetic relationship might easily be discredited as empirically trivial except in specific medical contexts. Via genealogy, affinity is established to people who are often negligibly connected, either on a genetic or cultural basis. In the present moment, a senator who might have become President reminded us of that by finding a very small component of Native American ancestry that might justify a claim to affinity that she may or may not have made. However, whatever the biologic meaning of miniscule fractions of DNA, tracing roots is alluring and sometimes a source of self-fulfillment, rather than self-congratulation or advantage. Attaching names and faces to remote ancestors teaches that we owe our existences to identifiable persons, as well as unnamed multitudes too easily overlooked. The genes shared on the basis of remote ancestral relationships may be few, but represent a connection that is real enough.

The world's great and lesser museums house millions of skeletons and artifacts of the dead, and our laboratories have trillions of their living cells. If an interesting person dies, scientists may race to analyze their DNA and rewrite the story of the origins or "true nature" of that person. On a microscale, competing teams of scientists proved that Anna Anderson was not Anastasia Romanov, as Anderson claimed throughout her life.

This book begins on a bit of a downer—death—but that is not where it ends. The dead are never quite that. Science rewrites the narrative of our connection to the dead, but more importantly teaches that we are connected to all life on earth, and to molecules essential to life. We share common origins, and to varying extents, nature, with each other and to the first self-replicating molecules. Science can help us embrace connections without the aid of mystical experience. Although we should question the conclusions they sell, credit is due to genealogists, scientists, and historians for unveiling hidden connections to others and to worlds of past and present. If others follow a spiritual, mystical, or religious path to this destination, I say good for them. In writing narratives of the dead, let us treat them and unloving artifacts derived from them with care and respect, however uneasy they may rest. The bell that tolls echoes in times to come.

Death

For most, "brain death" is simply an equivocal formula that they learned to accept during training, for no more profound a reason than that everyone else accepted it.
Western society seems to be rapidly approaching the stage where the moment of death will be determined not so much by objective bodily changes as by the philosophy of personhood of those in charge.

Neurologist D. Alan Shewmon, in his Apologia

The year of Terri Schiavo's death is disputed. In 2005, Schiavo's feeding tube was removed at her husband's request, made 10 years earlier. She died 13 days after "artificial" sustenance of her body was stopped. Or, in 2005 there was no "her," Terri Schiavo having died 15 years earlier at the age of 21 when she suffered a cardiac arrest and collapsed in her apartment.

Precious minutes passed before emergency personnel arrived and began trying to resuscitate her. In those minutes, and, as we now know, afterwards when reactive processes caused further loss, her brain was damaged and the young woman engaged with the camera was irrevocably gone. Less than 3 months later, Terri Schiavo was declared to be in a vegetative state. From 1990 to 2005 her body was, although with eyes open, something very different, and lesser—but was Terri Schiavo alive or dead? It is a raw paradox that lifesaving medical technologies have led to incalculable benefit and gratitude, but also to suffering, doubt, and conflict.

The benefits of modern shock/trauma and intensive care and longer-term management and rehabilitative medicine that enabled Terri Schiavo's body to be sustained can scarcely be overestimated, even if they are difficult to calculate in the first place. Medical miracles are not so easy to quantify. The frequency and cost are. Most relevant to what happened to Terry Schiavo are intensive care and long-term care of comatose patients. More than $108 billion dollars is spent yearly on critical care in the United States, about one in seven health-care dollars and 0.72% of gross domestic product (Society of Critical Care Medicine, https://www.sccm.org/Communications/Critical-Care-Statistics). Nowadays, anyone admitted to a hospital from an emergency room is more likely than not to be sent directly to an intensive care unit (ICU). Some six million Americans are cared for in ICUs every year. In ICUs, interventions that to previous generations would have been astounding are now delivered to 50,000 patients on a given day. Teams of intensivists sustain respiration, circulation, electrolyte and metabolic balance, and nutrition. They protect the brain, even extracting clots from cerebral arteries, and induce long-lasting comas to protect and heal

Immortal. https://doi.org/10.1016/B978-0-323-85692-8.00001-0

the brain. They partly replace functions of failing organs such as kidneys. Via invasive and noninvasive monitoring, their interventions are precisely targeted: for example, the use of brain scans to determine if stroke victims have clots suitable to extract.

The ICU is a place of life, but also of death in large measure. It frequently happens that after little delay ICU intensivists help life to end with dignity, in consideration of the wishes of patient and family, and in line with the limitations of medicine to sustain life or provide a pathway back to a life worth living. Depending on age and disease severity, between 1 in 10 and 3 in 10 patients admitted to the ICU die during an admission. The other side of the equation, although not the sort of question that is highly amenable or appropriate for controlled study, is that many of the other 50,000 patients/day would otherwise be dead except for the ICU. Personally, I would be twice dead except for life-saving care when I was in septic shock following pyelonephritis. Altogether, the lives of tens of thousands of victims of stroke, heart attacks, trauma, toxins, and sepsis are saved every year.

In America, thousands of people become comatose yearly, most as a result of traumatic brain injury, and most of those from motor vehicle accidents. As may sound oxymoronic, but is not, most accident victims who are comatose for several days will recover—coma from other causes, such as cancer and cardiac arrest, has a worse prognosis. Recovery is not all or none. Often, consciousness and other cognitive, emotional, motor, or sensory faculties remain, and these may last lifelong or recover only very slowly. Prognosis is predicted by type of injury and by how long the chronic vegetative state has persisted. For coma caused by traumatic brain injury, a year may elapse. For coma caused by anoxia, such as Terri Schiavo suffered, recovery seldom happens after 3 months.

I was one of a generation of medical students fortunate that neurologist Fred Plum and his colleagues, including Bryan Jennett, had recently (in 1972), defined coma and persistent vegetative state (PVS), as described in the textbook by Plum and Posner: *Diagnosis of Stupor and Coma*. The neurologists who taught us helped us to perceive that individual cases were part of a pattern and to understand what was happening inside the patient. At that time, almost everything known about the state of a patient's brain was inferred from clinical signs, symptoms, and responses. A decade later, Plum and David Levy performed the first functional brain imaging of coma, using fluorodeoxyglucose positron imaging tomography (PET). Techniques such as PET, but also the technically much simpler electroencephalogram (EEG) and magnetoencephalogram (MRI), are windows into the function and structure of the brain that help make sense of clinical signs and observations and help predict prognosis.

Incredibly, modern brain imaging studies reveal that even the most damaged brains retain functional networks performing fragments of the capacities of the brain prior to injury. Even this wreckage of intact function is far beyond human capacity to devise, and unique to the brain of the patient in coma and to their unique pattern of injury and healing. The problem, as shown by investigators working in several countries, including David

Menon in England and Steven Laureys in Belgium, and Nicholas Schiff in the United States, is that the circuits that remain functional are disconnected, preventing integrated response. Especially important to a return to integrative function is the upper brainstem, the so-called reticular activating system, and—also deep in the brain—regions of the thalamus. If these neural switchboards and pacemakers are irreversibly damaged, recovery is doubtful. Could the functions provided by these regions be replaced artificially? One method being studied is deep brain stimulation (DBS), a technique in which an electrode is placed in the brain and used to drive, or inhibit, neural activity. If restored, was the person declared dead really there all along, just waiting for the "kiss" of the right technology?

One of the first patients in whom DBS was attempted was Terri Schiavo. In a highly cited study published in *Nature* in 2007, Schiff and my friend, neurosurgeon Ali Rezai, reported that thalamic stimulation was able to recover some coordination of fragmented activity in a comatose patient (Schiff et al., 2007)—but not Schiavo, and not a success in restoring a patient from a PVS. So far, and despite open-label clinical trials in many patients, DBS has not shown lasting benefit in comatose patients, but could it be a matter of improving DBS technology, treating the right patients, or combining DBS with other interventions, including many new tools such as chemogenetics, optogenetics, and stem cells, unavailable a generation ago? It would be easiest, and perhaps kindest, to say "no" but that would not be forward thinking. Many modalities—for the most part crude and unfocused compared to what is coming—are being tried, including drugs such as amantadine and apomorphine, and external direct current stimulation. This is the not quite irrational hope for miracle cures that can sustain hope in the face of the message that a loved one is dead. As advocated by Nicholas Schiff, a pioneer in neuroimaging and DBS in comatose patients (Dana foundation website: https://www.dana.org/Cerebrum/2003/Hope_for_%E2%80%9CComatose%E2%80%9D_Patients/), the possibility that DBS can enable a patient in PVS to recover should not impinge on a person's right to die. But might it not? How many would agree that there is a right to die from appendicitis or from an acute blood loss that can be rescued by transfusion? The latter is exactly the dilemma faced in treating some children of Jehovah's Witnesses. Technology has always defined the boundaries and goalposts of what is morally acceptable, and it is true in both medicine and sports that the rules of the game are changed if players suddenly become unstoppable, which could be a metaphor for people learning to extend life by a century or two.

Since 1972, many icons of medicine have been broken, and it is a compliment to be called an iconoclast, but Fred Plum's conception of the natural history of stupor and coma, developed with his residents Jerome Posner, August Swanson, and Donald McNealy, endures.

From the Fourth Edition of *Stupor and Coma*: *"Coma, from the Greek 'deep sleep or trance,' is a state of unresponsiveness in which the patient lies with eyes closed and cannot be aroused to respond appropriately to stimuli even with vigorous stimulation. The patient may grimace in response to painful stimuli and limbs may demonstrate stereotyped withdrawal responses, but the patient does not make localizing responses or discrete defensive movements. As coma deepens, the responsiveness of the patient, even to painful stimuli, may diminish or disappear."*

Sleep is not coma, being reversible by stimulation. Sleeplike coma rarely lasts more than 2–3 weeks, being followed by a vegetative state originally described by Jennett and Plum in 1972. In that state there is limited recovery of arousal, with alternating periods of an eyes-open state mimicking wakefulness and an eyes-closed state that is an imitation of sleep. In a vegetative state, the patient has no awareness or response even when eyes are opened, and if this persists more than a month, the vegetative state is called persistent vegetative state (PVS). Rarely, patients have "locked-in" syndrome, which can be diagnosed by careful clinical exam, detecting, for example, eye movements that are still possible, and assisted by technologies such as EEG and MRI.

The "modern" dilemma is that the same technologies can sustain the body essentially indefinitely in a vegetative state, waiting for some future cure that may never happen. Can this be immortality? When it all goes wrong—which is our fate as humans—and in cases where one family member or another does not accept that brain death is death, it often leads to months or years of exorbitantly expensive medical/mortuary (depending on one's view) care, years of litigation, politicization at the highest levels, and Job-like, nearly 24 hours a day suffering for surviving relatives.

As freshman medical students, a small group of us were shown a video case history of a young man who had suffered severe burns. Sometimes we interviewed simulated patients, but in the film clip this man's face was scarred, and when he said he wanted to die we took that as fact. Burns I knew anyway, because one of my father's many roles was as pediatric immunologist at a Shriners Hospital for burns. Talk about nepotism, I later was night clinical lab tech there doing toe sticks on burned children and measuring blood indices. Also, bringing me a little closer to a burn victim's experience, I accidently boiled off the skin of the top of my left foot with a pressure steamer while working on the Galveston wharves. No big deal—some of my coworkers had missing digits. After benefiting from some of the advanced procedures of the time, including months of ripping off eschar with a hemostat, my burn eventually healed, but I could slightly better imagine what a young man burned over 90% of his body endured, and I would hazard a guess that others who have had major burns would agree, one's bad experience providing cover to make such claims, whether or not it is true.

Our mentor was Melvin Schreiber, a master radiologist. He asked if we would also wish to die under those circumstances. Decades later, I was able to personally thank him for influencing me, as my father did, to keep medicine personal, but to base it on data and diagnosis, rather than anecdote, test result, or symptom. After being burned, the young man had endured years of painful treatment and rehabilitation. Like the badly burned kids I had seen, almost all his skin was gone. His shoes protected his feet, but his clothes and hair had ignited, burning almost every other surface. He was blind and in pain. More surgeries were coming. The life he knew was gone, and he did not accept the life he faced. So what was the vote? I would be confabulating if I stated the exact tally, but according to the advertising of the time, we were the Pepsi Generation. The vote was about 10 to 1 for assisted death. I said that I could never be sure what I would do if I were in his shoes (and my feet had all the skin left unburned). But I was mostly sure I would still want to live—to hear, to smell, to taste, to talk, to think, to get better, day-by-day, at blindfold chess, to argue when life begins and ends, and what all this suffering means, if anything at all. These are gifts. Now I am more sure than ever that I do not know, but each and every dive further down the one-way tunnel of aging and infirmity has taught me to appreciate what I have rather than dwell on the losses. Being honest, I know I will not be as brave, but daily examples of patients and acquaintances who have lost more than me—a breast, a limb, vision, and even part of their brain and mind—and still found reasons to live sustain me, and I see what their examples have meant to others. But there must be some point where enough is enough and, disconcertingly, it is now very often someone else's job to decide when that time has come, not the person who cannot speak, or think, for themselves.

Disturbingly, a patient who is brain dead can exhibit all kinds of other signs associated with life: breathing, maintaining cardiovascular function, thermoregulating by showing gooseflesh and shivering, having endocrine function, menstruating, displaying jaw and snout reflexes, jerking of the limbs, and spasms. Only 3 weeks after a coma from which a person will never recover, the eyes open, as can be seen in the photo of Terri Schiavo, taken 15 years after she lost all awareness. As any intensivist keeping patients alive in an ICU knows, more of the body's essential functions—immunity, digestion, waste, disposal, respiration, synthesis of essential blood proteins—are carried out by organs located below the neck rather than by the brain. As one ICU intensivist joining Alan Shewmon in the care of a brain-dead patient exclaimed, "Isn't it amazing how well the body can function without a brain!" Even worse in the eyes of observers, more complex movements are also possible, even including the arms crossing the chest and sitting up, the so-called "Lazarus sign" (Reviewed by Shewmon, 1997).

Writing about Terri Schiavo, bioethicist Arthur Caplan observed how preoccupied, polarized, and puzzled we are by the question of when she died, the definition of death, and who gets to decide for a person who can no longer act on their own behalf (https://www.nbcnews.com/health/health-news/bioethicist-tk-n333536). Is it up to spouse, parent, or child to decide? Someone with power of attorney, or a court-appointed guardian? An omsbudsperson? What happens when they do not agree? Do we defer to a physician, or a panel of physicians and bioethicists acting on behalf of the hospital? How about a "death panel," a private or government health administrator, an insurance company, a judge, a governor, or a group of senators?

To resolve what was essentially an intrafamily dispute, the Florida legislature passed "Terri's Law" and Governor Jeb Bush intervened to keep her body alive, although courts had repeatedly sided with Terri Schiavo's husband Michael Schiavo. Law enforcement officers removed Schiavo from the hospice where her feeding had been stopped and placed her in another facility where a feeding tube was surgically inserted. Then, she was returned to the hospice. Eventually, Terri's Law was struck down as unconstitutional, the Supreme Court refusing to intervene.

In the mid-1990s Senator/surgeon Bill Frist was one of the most powerful men in America and perhaps on track to become its 43rd president. Dr. Frist's hopes foundered after, and in part because, he staked his career on Schiavo's case, despite never having examined her. When it became clear that lower courts would withdraw the tube feeding, Congress passed an emergency law transferring the decision to Federal courts about an hour after midnight on March 21, 2005, and the law was signed a few minutes later by President George Bush, who had flown in from Texas especially for that purpose.

"Higher authorities" weighed in. Pope John Paul instructed Catholics, and Catholic institutions, that water and food could not be withheld from bodies in a vegetative state. Did they not still have a soul? There being no data on where in the body the soul is located, how would one know? The Pope's opinion gained relevance when Schiavo's parents argued in court that Terri was a devout Roman Catholic who would not wish to violate the Church's teachings.

The living dead: In many end-of-life cases, including Terri Schiavo's, expert neurologists testify that the brain-dead person is a "corpse." At Schiavo's autopsy, her brain was found to have shrunk by half. With limits on capacities for neural regeneration and functional recovery, she would never have recovered beyond the capacities of an animated corpse, at least not via technologies known or on the horizon. With components of her brain or body—single-induced pluripotent stem cells even, or a bioengineered living computer transplanted into her brain—it cannot be denied that one day a living, breathing, socially interacting human could be made, and it might look like Terri Schiavo. But it would not *be* Terri Schiavo. The Terri Schiavo who family and friends knew, and that knew herself, was irretrievably gone. But, was she a corpse?

Brain death is today not the definition of death in two US states and some countries, and until the 1960's cardiorespiratory failure was, as investigative journalist Rachel Aviv put it, "the only way to die." Death had to be redefined because advances in cardiorespiratory support and other ancillary care made it possible to keep the heart beating and the body alive, even if the person would never wake up. In 1967, Harvard bioethicist Henry Beecher noted, "Permanently comatose patients, maintained by mechanical ventilators, were increasing in numbers over the land…" Beecher acted, organizing a committee of acquaintances: one historian, one lawyer, one theologian, and ten doctors. In the following year, these 13 men outlined the brain-death standard in a highly influential article in the *Journal of the American Medical Association* (Beecher et al., 1968). They justified their standard in two ways, neither of which directly addressed whether the brain-dead person was dead:

(1) The cardiorespiratory care of brain-dead individuals was futile, and expensive both for hospitals and families.
(2) Obsolete criteria for brain death were creating problems for organ transplantation, a burgeoning field of medicine that was saving thousands of lives and restoring thousands to near full health.

Beecher and his colleagues realized that, whereas a decision to take a comatose person's organs had to be taken in a moment, death was not a moment, explaining that: (1) death is a process, not an event; (2) where to draw the line along this process for legal purposes is arbitrary and culturally relative; and (3) the utilitarian motivation of saving lives through organ transplantation is a good (and sufficient) reason for drawing the line at "brain death" [Beecher 1978; Beecher and Dorr 1971]. Later, these nuances were blurred, and still later the brain-death standard was itself reevaluated by Shewmon, as will be seen.

In 1981, a Presidential Commission developed a theory of death, defining death as the moment when the body stops working as an "integrated whole." This formula was embraced by the American Medical Association, and most US states quickly adopted this brain-death standard. In 2018 only two states, New York and New Jersey, allowed religious exceptions. David Wikler, the philosopher on the Commission, is quoted by Rachel Aviv as saying, "I thought it was demonstrably untrue, but so what?" Wikler had argued for the narrower but still problematic formulation of death based on integrity of cerebral function, like the persistent vegetative state (PVS) standard of the American Academy of Neurology. And exactly what is cerebral function? The problem, as the Commission's chair, Edmund Pellegrino, said in a note appended to the report, is that definitions of life and death are self-recursive, consisting "in some form of circular reasoning—demarcating death in terms of life and life in terms of death—without a true definition of one or the other."

Legally, but also out of empathy for those facing end of life and for their families, a standard for death was essential. Due to advances in medicine, the ability to extend life, and suffering, is untold, and open-ended. The definition of death hangs over the involuntary termination of life support, as well as any discussion of right to die. Unfortunately, the definition of death as brain death, essential as it was, from the beginning was flawed. Why? Because human life itself has never been adequately defined, and probably cannot be, life being a continuing and never-ending cycle. Medical advances require new standards both for the beginning and end of life, but as philosopher Peter Singer has written, the brain death standard is "an ethical choice masquerading as a medical fact" (quoted by Rachel Aviv).

Arguing for their even broader and debatably more poorly focused "integrated whole" definition of life and the transition to death, the Presidential Commission claimed that the organs and subsystems of an artificially supported body would inevitably degrade. Glibly, and wrongly, they claimed the "heart usually stops beating in two to ten days." Except when care is better, and then it does not stop beating. Often, rescue from cardiorespiratory death is prelude to the now well-recognized phenomenon of PVS. By 1989 the American Academy of Neurology clarified that PVS (a) dooms the patient to dismal prognosis for recovery of higher functions, and (b) is otherwise compatible with other functions of the body: *The persistent vegetative state is a form of eyes-open permanent unconsciousness in which the patient has periods of wakefulness and physiological sleep/wake cycles, but at no time is the patient aware of him- or herself or the environment. Neurologically, being awake but unaware is the result of a functioning brainstem and the total loss of cerebral cortical functioning ....Persistent vegetative state patients do not have the capacity to experience pain or suffering. Pain and suffering are attributes of consciousness requiring cerebral cortical functioning, and patients who are permanently and completely unconscious cannot experience these symptoms [American Academy of Neurology 1989].*

Whether or not pain and suffering require consciousness is a question unresolved, but insistence on the primary importance of a functioning cerebral cortex can take one to strange, untenable places, as in the 1990 report from the Medical Task Force on Anencephaly, a condition in which the infant is born without a cerebral cortex:

Experience with other cerebral lesions indicates that the suffering associated with noxious stimuli (pain) is a cerebral interpretation of the stimuli: therefore, infants with anencephaly presumably cannot suffer. Anesthetic agents may eliminate the subcortical responses to noxious stimuli but are not necessary to minimize or prevent suffering. [Medical Task Force on Anencephaly 1990 (pp. 671, 672)].

About two decades after the Presidential Commission, neurologist D. Alan Shewmon reignited the controversy, having studied 175 patients who had survived for months or years in PVS (Shewmon, 1997, *Recovery from "Brain Death": A neurologist's apologia*). Shewmon had carefully prepared his critique of brain death, having anticipated the likely reception: "In professional circles, dissenters from the 'brain death' concept are typically dismissed condescendingly as simpletons, religious zealots or pro-life fanatics." However, the questions raised by Shewmon and other neurologists such as Calixto Machado (Machado and Shewmon, 2004) eventually led to a new Presidential Commission. This committee's 2008 report "Brain Death" abandoned the idea that the brain was necessary for the body to function as an integrated whole, embracing the brain-death standard wholeheartedly.

This was not the result Shewmon would have wanted. In his *Apologia* Shewmon had rejected whole brain death and cortical death: "'brain death,' in any and all of its neuroanatomical and semantic variations, is really not death after all but rather a state of deep and irreversible unconsciousness in a critically lesioned but still live patient." In many cultures there are other ways of defining life than brain function, but as Shewmon observed, it was probably inevitable that neurologists, as well as other intellectuals who value above all the outputs of the mind, would define death neurologically, the body being the "carrying case" for the brain. But however one defines death, would one dissect, bury, or cremate a body that was still breathing? Or, how comfortable are we with definitions of death that rely on loss of personhood, when these take us perilously close to, or directly into, the domain of the Nazi concept of biologically alive nonpersons, popularized in the book *Permission to Destroy Life Unworthy of Living*, written by German psychiatrist Alfred Hoche and jurist Karl Binding and published in 1920.

The one time the Supreme Court ruled on withholding of care in PVS was in 1990, when in the case of Nancy Cruzan versus Director, Missouri Department of Health, the Justices split 5–4, ruling in favor of Missouri that care could not be withheld in the absence of "clear and convincing" evidence for what Cruzan's expressed wishes were. Chief Justice William Rehnquist wrote the opinion, which had several important holdings, including: "The United States Constitution does not forbid Missouri to require that evidence of an incompetent's wishes as to the withdrawal of life-sustaining treatment be proved by clear and convincing evidence. "The Court also ruled that a family's decision-making need not be accepted as a substitute for the state's. Although Rehnquist wrote the opinion for the majority, five of the other eight justices wrote separate opinions. Justice Antonin Scalia wrote that a patient did not have a right to assisted suicide, by starvation or other means. By that time, Cruzan had already been in PVS for 7 years, ever since she had been thrown from her car and found facedown in a water-filled ditch. In other words, to the Supreme Court, in a precedent that has not been overturned to this day, 30 years later and after so much has changed in medicine, a brain-dead person has more legal status than does a corpse. Within a few months, the Cruzans had gathered more evidence that Nancy Cruzan would have wanted treatment withheld, the State of Missouri withdrew its opposition, and Nancy Cruzan's body stopped functioning the day after Christmas, 1990.

A decade before Nancy Cruzan's accident, Karen Quinlan entered PVS after overdosing on alcohol and Quaaludes at the age of only 21 years. Perhaps more significantly, Cruzan—like Terri Schiavo later—was crash dieting, Quinlan not having eaten for 2 days so that she could fit into a certain dress. That same year, Quinlan's parents, joined by the hospital, which faced potential homicide charges, sued to have Karen Quinlan taken off her respirator. In line with a papal declaration, Quinlan's parents considered artificial ventilation, but not tube feedings, to represent "extraordinary means" of preserving life. In 1976, the New Jersey Supreme Court ruled in the Quinlan's favor and the respirator was removed. However, with the tube feedings continuing, Karen Quinlan "lived" for another 9 years, until 1985. Her body was 65 pounds at time of burial. Quinlan's parents, Joseph and Julia, and later Julia, remembered their daughter, and their struggle, in two books.

To the family, the brain-dead loved one does not look or act like a corpse. Often, after some weeks or months, the eyes open, the body reacts with primitive reflexes, and random movements begin. With almost all cortical pyramidal cells dead, as actually had happened in Schiavo's case, the intact brainstem and spinal cord maintain cardiovascular function and the abilities to grasp, swallow, and make random movements. Confronted by a breathing, moving, reflexively reacting shell of a person, it is natural for family members to hold on dearly. Nothing in millions of years of brain evolution, during which time comatose people quickly died, could have prepared our brains to reject a daughter whose eyes we can look into in favor of neurologists' assurances and brain scans and EEGs showing absence of brain function.

Postmortem: Terri Schiavo is remembered in legal and biomedical history, and in books written both by the husband and family, for inspiring and perhaps illuminating the debate on the boundary between life and death. Less well known is that Schiavo's death drew attention to our responsibility for the well-being of the living. Schiavo was already suffering and in danger of death for an entirely preventable reason: bulimia. In 1992, Michael Schiavo, on behalf of Terri Schiavo, was awarded $6.8 million (later reduced to $2 million) in a malpractice case because doctors had failed to recognize bulimia as a cause of Terri Schiavo's infertility. Unfortunately, concern over who might inherit the money that Michael Schiavo had set aside in trust for Terri Schiavo clouded later litigation and polemics on whether her tube feeding should have been stopped.

Cardiovascular collapse in a 27-year-old woman is unusual. The cause of Schiavo's cardiovascular collapse was excessive dieting leading to very low levels of potassium. The heart depends on an ionic balance between extracellular and intracellular compartments, and in that balance potassium levels are key. Terri Schiavo's serum potassium level, 2 mEq/L, was low enough to produce fatal arrhythmias, the lower end of the normal range being about 3.5 mEq/L. Her physician wrote in the medical record that she "apparently has been trying to keep her weight down with dieting by herself, drinking liquids most of the time during the day and drinking about 10–15 glasses of iced tea" [USA Today, 2005]. Friends observed that she frequently disappeared after meals, probably to vomit. Like many women with eating disorders, she stopped menstruating. All this was because she had once been an unattractive 200-lb teenager who wanted to be attractive, and found a way of making herself so, losing 65 lb. Gary Fox, the lawyer for Terri Schiavo's husband, correctly observed, "While there is no cure for bulimia, there were things that could and should have been done for her that would have controlled it" (https://usatoday30.usatoday.com/news/health/2005-02-25-schiavo-eating-disorder_x.htm). Bizarrely and unnecessarily, after Terri Schiavo's death, the medical examiner ruled that there was no evidence of bulimia. It is doubtful that evidence of bulimia would have endured after Schiavo had been fed by tube for 15 years. Representing the views of the family, Schiavo's brother, Bobby Schindler, Jr., said, "The fact that the medical examiner ruled out bulimia and ruled out a heart attack, without a doubt, adds more questions."

Jahi McMath, who was Black, changed the public face of PVS. It is no longer only young, White, and female. It is potentially any of us. Furthermore, Jahi's case inverted the struggle between family and state. Rather than suing for an end to the suffering, her family struggled against all odds to keep Jahi, or Jahi's body, alive.

Journalist Rachel Aviv (*The New Yorker*, 2018) delivered a penetrating portrait of Jahi and the family's enduring love. Up to a certain point, the facts are mundane. Jahi McMath did not suffer from bulimia, but like Karen Quinlan and Nancy Cruzan, her coma can also be traced to weight problems in a young woman. An obese 13-year-old, Jahi had problems breathing while sleeping. She snored and suffered from sleep deprivation and social embarrassment. All of this is consistent with the harmful-to-brain syndrome known as sleep apnea. Tonsillectomy was recommended, and Jahi's mother Nailah overrode the child's natural reluctance, telling her that, freed of the snoring and sleep deficits, she would have a better life.

Jahi McMath's operation, a 4-h affair, was as usual, in the world of surgery, "successful," although the surgeon had noted that one of the girl's carotid arteries was anomalously close to her pharynx, increasing the risk of the procedure. The tonsils were removed, but only an hour afterward she began spitting up blood, and over the course of the next few hours this continued in ominously large amounts, her blood filling a bucket. The family's pleas for attention were mainly ignored. Soon after midnight a family member who happened to be a nurse alerted staff that Jahi's blood oxygen saturation had dropped. Doctors and nurses rushed to intubate Jahi, but soon after, a doctor said," Oh, shit, her heart stopped." She had also stopped breathing.

A person whose heart and breathing have stopped is regarded dead by any rescuer administering cardiopulmonary resuscitation (CPR). The code team trying to bring the victim back from beyond the divide between the living and the dead knew well that CPR administered perfectly is at best partly effective. The odds of surviving a cardiac arrest are highest when a person collapses in a hospital in front of medical personnel. Jahi collapsed in an intensive care unit—the best of scenarios for intervention, but also a place where the sickest patients are found. Even in the best of circumstances, and despite the application of medicine's most advanced interventions by highly trained personnel, the odds are only 1 in 5 for the average patient surviving after CPR (Brady et al., 2011). In the hospital, and as may have happened in Jahi's case, doctors are able about half the time to restore spontaneous circulation—the heart resumes beating and pumping blood to the lungs, brain, and other organs. However, life cannot be maintained if the underlying cause is uncorrected or if too much time elapsed before the brain and other vital organs were reperfused. Jahi was bleeding and during the hours of CPR had already suffered damage to her brain and other organs.

As reported by Aviv, 2 days later Jahi's bleeding had stopped, her heart was beating, and exchange of gases through her lungs was being successfully maintained with assistance of a ventilator. However, she was, under California law, dead, or as the family claimed the hospital's chief medical officer said, "She's dead, dead, dead," while pounding his fist on the table.

Whatever is the truth about the words spoken, Jahi McMath's test results are not in dispute. Her electroencephalogram was flat. She had no gag or vestibular reflexes. She could not breathe on her own. Later, a sophisticated neuroimaging test showed that her brain had no, or very low, metabolic activity. Time passed, first an interval of several days when the hospital allowed the family to gather and come to terms with their loss. Sadly, by a week after the surgery, the family had rejected the hospital's verdict. Instead of a dead body, they saw a living child whose skin was warm and soft, and whose limbs moved. Hurtfully, they also saw a neglectful and racist hospital that wanted to "kill her off," harvest Jahi's organs, and reduce their medical malpractice exposure— California having a cap of a quarter of a million dollars for pain and suffering of the dead, but no limit for the living.

The medical ethics committee of Oakland Children's Hospital rejected the family's request for a tracheotomy to facilitate Jahi's breathing: "No conceivable goal of medicine – preserving life, curing disease, restoring function, alleviating suffering – can be achieved by continuing to ventilate and artificially support a deceased patient." On the other hand, the committee noted the "tremendous moral distress" of personnel caring for Jahi, and "significant concerns for justice and fairness," given the resources being allocated to care for a child already dead. Meanwhile, Jahi's body was not being fed, tube-feeding not having been initiated for the same reason.

The family enlisted a pro bono lawyer, publicized Jahi's case on social media, and raised money to fund Jahi's transfer and "life support" at a cooperative facility. Jahi was declared dead and her body released to the family. Two states—New York being one of them, allow religious exceptions to the brain death rule. Jahi's body was flown cross-country to the other, New Jersey, where the body was cared for at a Catholic hospital. The following year it was discovered that Jahi's body had recovered the ability to breathe on its own, and the family was able to move it to a New Jersey apartment, where they cared for it with home nursing assistance until 2018, when she died. The medical malpractice lawsuit lived on. The hospital's lawyers—echoing a ruling by the IRS that Jahi could not be declared a dependent, argue that a dead body, a "corpse" as their expert testified, could have no legal standing and that there was no logic to holding the hospital liable for maintaining it. In 2018, 4 years after her surgery, Jahi McMath was declared dead. Her mother, Nailah Winkfield, and stepfather Marvin were at her side in a New Jersey hospital. The cause of death was listed as bleeding consequent to liver failure. By a judge's order, a medical malpractice lawsuit, now a wrongful death suit, continued. Barring a settlement, in a federal civil rights lawsuit a jury would have to decide whether Jahi was alive or dead, reversing the original death certificate. In a self-contradictory and vastly oversimplified statement, the spokesperson for Oakland Children's Hospital, Sam Singer, told the *San Francisco Chronicle* he was relieved to see Jahi "finally" rest in peace. "Her case created a national debate over something that unfortunately people don't understand. And it's very simple: Dead is dead. You don't come back from it." Afterwards, Nailah Winkfield spoke for the family: "My daughter died on June 22nd, 2018, not December 12, 2013. Jahi McMath was not brain dead or any other kind of dead. She was a little girl who deserved to be cared for and protected, not called a dead body… Jahi has forced the world to rethink the issue of brain death. My every day was focused and revolved around Jahi. I loved seeing her every morning and kissing her goodnight every night. The hole in my heart left by her passing is huge" (Reporting by Samantha Schmidt, *The Washington Post*, June 29, 2018).

In his *Apologia*, and earlier in an article in *Thomist*, Shewmon—who reviewed many videos of Jahi McGrath and has said he did not believe she was dead—suggests a thought experiment for what parts or functions of the body define life and death. The head of a person is cut off, as happens in certain countries, but physicians immediately intervene to sustain both the head and the body below the neck. This idea is not as far-fetched as it sounds. Heads of deliberately aborted 5 months gestation human fetuses were sustained for several hours (Adam et al., 1973). Also, we can infer that most people who contract to cryogenically freeze their own heads or those of relatives have already answered the question that Shewmon then asked: did the self of the person who was decapitated reside (a) in the head, (b) in the body, (c) both, or (d) neither? Having answered Shewmon's question, for example, with a) the head, one can then ask, wherein is the person if we section the brain into its two hemispheres? And so on. What if we are left only with cells from which something very similar to a person might be cloned? This is where Shewmon's thought experiment, which he years later concluded by deciding that both body and head were alive, runs up against a bit of a wall. Shewmon, a Catholic, concluded that the person's soul might reside in both the head and the body after the head was severed. But then, why not a donated kidney, blood, or a cell? As often happens, the answer is probably not at either extreme, and may be found in some other axis by which life might be defined.

The definition of death, and what to do about bodies in PVS, may not be settled any time soon because, as Pellegrino implied, we lack the frame of reference needed to define life and death. However, we do know the first principle of respect for people, and for people in states that we cannot fully comprehend. Informative though it would be for many purposes, we take organs from brain-dead people, as Jahi McMath's family feared, but not without permission, and we do not use them for medical experimentation. Nor are medical students allowed to dissect the bodies of the brain dead, or to practice a variety of medical procedures. However, medical science has license to experiment in diverse ways on cells made from that same nonconsenting person's body.

In his *Apologia*, Shewmon describes how stunned he was in 1989 to personally observe that hydranencephalic children born with no cerebral cortex could interact adaptively and socially. "For example, Andrew could scoot around the house on his back by pushing with his legs, without bumping into furniture; during the summer he would scoot through open doors onto the sun porch. He was obviously not only conscious but had at least rudimentary vision and voluntary motor functions." Andrew, and others like him who Shewmon examined, was a child of a lesser god, but something much more than an inanimate collection of the same number and type of atoms. If it is utilitarian to harvest the organs of a person whose status is in doubt but whose organs are needed, we align ourselves with the family who hopes, almost surely in vain, that their loved one will reawaken, or who values them as they are in PVS.

2 ◎

HeLa: The resurrection of Henrietta Lacks

So our memory is the only help left to them. They pass away into it, and if every deceased person is like someone who was murdered by the living, so he is also like someone whose life they must save, without knowing whether the effort will succeed.

Theodor Adorno

The most famous human cell line is HeLa, derived from the cervical cancer that killed Henrietta Lacks in 1951. As I finish this book, I am 68 years old, and HeLa is 69 and will always be one year older than me, assuming that my cells are also in culture. HeLa is a tool that I and other biologists have used through the course of our careers, but it is also a point of departure for uses to which people put derived pieces of others, and ethical limits on those uses and how we can derive things from people, even dead people.

Henrietta Lacks lives in public memory because of HeLa, and HeLa constitutes the main, if unwilling, legacy of her life. The disputes surrounding creation and use of HeLa are made tangible because the tissue used to derive the cell line was used without permission. Furthermore, it is extremely doubtful that if Lacks had consented that she would have understood to what she was agreeing. George Vey, the scientist who made the HeLa cell line, could have himself not anticipated the vast usage of HeLa. Vey made HeLa before Watson and Crick announced the structure of DNA and a half century before the human genome was sequenced. The genomic revolution enabled the HeLa genome to be sequenced and its biology manipulated in a myriad unforeseeable ways. Furthermore, Henrietta Lacks' individual and familial identity were not publicized by Vey. They were brought to the forefront of the public mind, or resurrected, by later controversies and by a book written 35 years later by Michael Gold, *A Conspiracy of Cells: One Woman's Immortal Legacy and the Medical Scandal It Caused*. Still later, and now in the internet age when nothing is forgotten, the public was reminded by a best-selling book by Rebecca Skloot, a journalist who immersed herself in Lacks' life and the story of HeLa.

As described in Skloot's *The Immortal Life of Henrietta Lacks* (2010), tons of HeLa cells have been grown. HeLa cells are in laboratories all around the world and used in tens of thousands of scientific studies and put to work for the production of lifesaving vaccines.

However, Skloot's book and other popular reports made a myth of HeLa. The myth is that Lacks's cervical tumor cells, and not others that could be readily derived from other victims with aggressive cancers of various kinds, and often have been, were essential.

Immortal. https://doi.org/10.1016/B978-0-323-85692-8.00002-2

15

That story is romantic fiction. HeLa "just so" happened to be the first cell line immortalized, and thus became a standard laboratory tool. By looking merely at how extensively HeLa is used, advocates such as Skloot confuse the value of HeLa, the cell, with the value of Henrietta Lacks, the person. They unfairly diminish the paramount contributions of scientists who used HeLa—which can as accurately be called Vey's cells—to make vaccines and discoveries.

It is widely known that Vey derived HeLa without the permission of Henrietta Lacks but seldom if ever acknowledged that advertisement of the relationship of HeLa to Henrietta Lacks and her surviving relatives was largely due to books and popular reports. That bell cannot be unrung. Also, many people concerned about the ethics of what was done would probably share my deep concern if they understand that if Vey had waited until after Lacks's "death" he could have done practically anything with and to her body under today's ethical guidelines. Research on dead people is not considered human research.

Much of the harm caused by the unauthorized taking and use, to whatever extent this harm is tangible, is attributable to persons other than Vey, who made HeLa at a time when informed consent was rarely obtained. For example, I used HeLa cells but was unaware that HeLa was shorthand for Henrietta Lacks. It was sufficient for me and colleagues, and we never discussed it, that HeLa was named after an otherwise anonymous woman who died of cervical cancer. Perhaps, as many scientists including myself thought, the donor was "Helen Larson." In either case, Lacks or Larson is only a name, and HeLa was effectively no woman or everywoman, given that scientists using it were not in the business of, or concerned with, hunting down details of the donor's life, as might have honored Henrietta Lacks, or tying HeLa and any findings derived from it to Henrietta Lacks's living relatives, as is unfair to them. Some people go through life worrying that they are being used (or exploited), but others worry about the alternative, which is to be useless. The concern that Henrietta Lacks was exploited by Vey is at odds with the claim that Henrietta Lacks made a unique contribution through having her cells taken. The first statement is truer than the second, because people should be allowed to choose if they want to be useful.

For better or worse, the controversy around HeLa changed things in a way that the act of making or using the cells did not. Perhaps for the better. A recent paper makes the far-fetched claim that HeLa is a new species, but it is less whimsical, or even true, that HeLa is an immortal, living extension of Henrietta Lacks, as well as the scientists who made and used it. Because of the controversy and disclosure, Henrietta Lacks's memory lives, as do her cells and the service to which they have been put.

There yet being no Blade Runner-type androids or Craig Venter-inspired artificial cells, HeLa or any human cells growing in an incubator also declare that a person existed. More than photographs or paintings, cells are living manifestations of the individual. Art can better reveal both superficial features and the unique inner essence of personality and changes in mood, but cells are more than an echo of a person, the DNA code contained within cells revealing secrets of the dead, and organoids, organs, or even whole persons potentially being grown from cells. The organ or whole organism is an authentic replicate of the original, potentially revealing its inner secrets.

What is the significance of a cell? The body, with its billions of interdependent cells, is a superficially seductive model of cooperative interaction. Free-living organisms may also play well with others but are also rugged individualists who stand well on their own against the physical environment, predators, and competitors, especially competitors of their own kind. Unlike complex free-living protozoans and less complex bacteria, cells functionally optimized to the shelter of the body are helpless in the hostile world outside. Except for a few specialized cells adapted for resilience in the external world, most, even HeLa, cannot survive outside the body for even a few minutes without special care and feeding. If the body's external barriers, immunoprotective systems, waste disposal, or delicate internal metabolic controls dysfunction, as happens in inherited and acquired diseases, and with aging, people sicken and die sooner or later. The specialized cells inside the body are quickly overwhelmed by single-celled, "primitive" invaders including bacteria, fungi, viruses, and protozoans of many different types. Or, cells stop functioning and die for lack of proper care and feeding. Failure of one cell—a cardiac pacemaker, a neuron controlling respiration, the endothelial cell maintaining vascular integrity—sets off a chain of other failures.

Biologists growing cells in culture try, in a limited way, to recreate the inner world of the body that French biologist Claude Bernard named in the mid-19th century. Bernard realized that this *milieu interior* was only maintained by a constant balance of "interior interactions and by compensatory reactions against environmental perturbation," saying, "The constancy of the internal environment is the condition for free and independent life...All the vital mechanisms, however varied they might be, always have one purpose, that of maintaining the integrity of the conditions of life within the internal environment." *Lectures on the Phenomena of Life Common to Animals and Vegetables* (translated by Hoff, Guillemin, and Guillemin L.), and via David Goldstein, 2009 (http://www.brainimmune.com). Early in the 20th century, American physiologist Walter Cannon coined the term *homeostasis* to describe the ballet by which internal balance is maintained. Cannon showed how one homeostatic hormone, adrenaline, triggers diverse metabolic and behavioral responses in reaction to threat or stress. Some would say that homeostasis—so well optimized in living systems—is a hallmark of life, but in this book one can find multiple examples refuting that, from the happily unfrozen tardigrade to the epigenetically mediated shifts in expression of whole cassettes of genes, and to nonliving systems that tend to restore balance to themselves, and, for example, the earth's climate.

Not until 1988 did Sterling and Eyer coin *allostasis*, realizing that the body often achieves stability through change, finding alternative (*allo* meaning other) set points better suited to new conditions. For example, global warming allowing, in winter we may don "gay apparel" and go caroling in hats, coats, and scarves—brightly colored to make it easier for others to see us in a snowstorm, or snowdrift. This is allostatic adaptation, because those thick, ostentatious clothes would have unnaturally drawn the attention of predators and hindered escape. The singing does not help. The discussion of allostasis and homeostasis here is brief, because other books have been written about the difference (Robert

Sapolsky, *Why Zebras Don't Get Ulcers*). Bruce McEwen, at the Rockefeller University until his death last year, identified long-lasting changes in the body's hypothalamic-pituitary-adrenal (HPA) axis as key in allostatic adaptation to stress, and introduced the concept of allostatic load. In the snowstorm example, the person adapts but is now encumbered by the weight and bulk of gear, and even the bright colors of winter clothing may be out of place in some settings. Allostatic adaptation often comes at a price: for example, addiction is frequently a long-lasting or point-of-no-return allostatic adaptation enabling the brain to deal with bouts of exposure to a psychoactive drug, but in the long run impairing behavior and emotion, and leaving the addicted person more vulnerable to stress or reexposure to the addictive agent (see Koob and Le Moal, and also Barr and Goldman, "Restoring the Addicted Brain," *New England Journal of Medicine*).

In the laboratory, cells can be shielded by artificial barriers and the proper care and feeding of most types of cells can be assured by culturing them under special conditions to mimic some of the vital aspects of Bernard's *milieu interior*. At the least, cell biologists approximate the optimal levels of salts, nutrients, vitamins, and temperature, and shield cells from competitors and physical injury. Certain cells, especially ones such as HeLa derived from tumors, are the most avidly growing and resilient. However, even these cells have to be carefully fed and maintained in a narrow range of temperature and pH (acidity/basicity). Most importantly, they must be shielded from free-living bacteria and fungi that would quickly overwhelm them. If one were to release several million "resilient" HeLa cells into a garden pond, they would die or be consumed within an hour. In the laboratory, HeLa and other cells are nurtured in a pH-buffered media broth containing essential nutrients and often luxuriously supplemented serum from the blood of fetal calves. The cells are grown in sterile flasks, usually at body temperature and carefully handled in a laminar flow cabinet, or "hood," whose curtain of air protects people outside from what is inside and protects the cells within from what is outside the hood.

In the early years of research on human cell biology, HeLa was a boon, mainly because it was the first of many immortalized cell lines to be made (for example, my lab has made immortalized cell lines from thousands of people), and because HeLa cells are relatively easy to grow. Instead of petering out after 70 or so cycles of cell division, an immortalized cell keeps going, and can be studied in genetically modified and unmodified versions by labs anywhere. It is less well advertised and is misunderstood that HeLa is hardly unique in its properties as a fast-growing, immortal cell, and that HeLa also played a passive role in a story of scientific setbacks. As discovered by gene sleuths, including Steve O'Brien (now director of the Dobzhansky Institute, St. Petersburg, Russia) at the National Cancer Institute, HeLa contaminated and overgrew numerous precious cell lines. Researchers who thought they were studying cells derived from the kidney or some other organ were misled because they were actually studying HeLa, a highly undifferentiated cervical cancer cell. It is comparatively easy to culture vigorous, fast-growing cell lines from highly undifferentiated tumors, and thousands of people succumb to such cancers annually. Thus the benefits or detriments of HeLa were not intrinsic to Henrietta Lacks but created by its users, and by how HeLa was used.

Although in keeping with biomedical research practices of 1951, taking of Henrietta Lacks's tissue without consent for biomedical research was a breach of ethics, and the very wide use of her cells, for example in DNA sequencing studies, amplified that original sin. Lacks's cells were not simply taken and forgotten, as happened in many other cases. The taking of the cells, and the uses and publicity that followed, illustrate wherein resides locus of consent—not in the tumor or derived cells, with which we can do whatever we want, but in the person, and if that person is unavailable, the next best thing may be to ask people related to that person. A compromise announced by the National Institutes of Health in 2013 to redress the error of the unconsented taking of Henrietta Lacks's cells was probably the best that could be achieved under the circumstances. Under the agreement, two relatives of Henrietta Lacks were included on a board evaluating researchers' requests for access to genomic information derived from HeLa. Lacks's DNA sequence also has implications for her living relatives, so inclusion of blood kin on the evaluation committee was appropriate. Although retroactive, this family-based group consent partly ameliorated the original ethical misstep. In the sad case of Henrietta Lacks, the wrong could not be righted, but the family was at least retroactively involved, and without monetizing what was never really contributed by HeLa, or attempting to compensate the family financially for a wrong that was done to Henrietta Lacks. Unfortunately, no living relative is really able to speak for the dead, and indeed other relatives later came forward to pursue legal recourse, acting on the false premise that HeLa had unique value. They sued Johns Hopkins University, where Lacks's cells were originally taken.

We cannot ask the dead to consent. For a person's cells or organs, either of which might be donated, the locus of autonomy and decision-making should be the person. When we use any animal, or human unable to consent—for example, because a human infant's brain has not developed sufficiently—it is our duty to act beneficently, minimizing pain and suffering and maximizing good that there may be in the world. However, overriding the principle of beneficence is the imperative to treat any locus of moral autonomy with dignity and respect. We ask the person because anything done with their cells has implications for them. It is that person's decision as to how to weigh the risks and the value of the beneficent outcomes we may predict, and not ours. We can educate but, in humility, remember that scientists could not anticipate the uses and misuses to which cells such as HeLa could be put.

Even in death we also respect peoples' memories and their wishes, understanding that the dead leave a legacy. Most people feel it is impolitic, and impolite, to speak ill of the dead and after life to uncover controversial personal details the dead can no longer dispute. Human cell lines are invaluable and even essential for genetic studies, and cells taken from people and grown in the laboratory are increasingly useful for cell-based therapies. In neurogenetics, we make numerous human cell lines. The cell lines enable us to take advantage of the fact that genes expressed in brain are often expressed in a blood or skin cell, and when there is a genetic difference altering brain function there is also an echo of that functional difference in a fibroblast cultured from the skin or a lymphocytic white blood cell cultured from the blood. Over the years, one scientist in my laboratory

(Longina Akhtar) made hundreds of mortal cell lines from skin fibroblasts and thousands of immortal lymphoblastoid cell lines from blood B lymphocytes. Some laboratories have made hundreds of thousands of immortal human cell lines, and these cell lines can be accessed via repositories, or "cell banks," such as the National Institute of General Medical Sciences (NIGMS) cell repository. The cell lines are invaluable for genetic studies, for example to understand how a specific genetic change alters function or to identify intrinsic biochemical correlates of disease. I described some of these stories of discovery of functional variants that alter human behavior in my book *Our Genes, Our Choices*.

A second step in many human studies (the first step being informed consent) is to obtain a blood sample. From that blood, cells can be temporarily cultivated or a cell line can be made, either a "mortal" one that grows a limited number of generations before petering out, or a cell line immortalized with a virus, or a "mortal" cell line grown indefinitely with the help of growth factors. Specialized cells of the skin or blood have undergone thousands of epigenetic changes altering gene expression and leading to specialized abilities. However, being epigenetic, these changes are reversible, if one can find the key to unlock the door back to pluripotency. In 2006, Nobel laureate Shinya Yamanaka made the startling discovery that temporary expression of only four genes, *Sox2*, *Oct4*, *Klf4*, and *c-Myc*, was sufficient to dedifferentiate specialized cells such as blood lymphocytes to immortal, pluripotent stem cells. These induced pluripotent stem cells (iPSCs) can be differentiated into mature cells of practically any type, including neurons, opening the way to biological studies in the laboratory.

In recent years, embryonic human and animal cells and iPSCs have been used to make *organoids*, capturing many of the structural and functional characteristics of organs, including liver, lung, brain, teeth, retina, and intestine, and enabling scientists to probe how these organs develop. Even some of the earliest stages of development, when cells self-organize into an embryo, have been partly replicated in tissue culture. Much remains to be learned to use these technologies safely and effectively, but within the next decade, the iPSCs, and cells and organoids derived from them, are likely to be taken from the laboratory bench (the incubator and tissue culture cabinet) to the bedside of clinical medicine. The transplantation of a person's own cells is potentially curative for a variety of intractable diseases, including Parkinson's disease, liver failure, Alzheimer's disease, and diabetes, which are in part caused by loss of particular cells or which may be caused by some deficiency that is reparable by gene editing.

Already, several types of cancer, including lymphomas and melanoma, are being brought to remission using cell therapies pioneered by Steven Rosenberg at the NIH. In these cell therapies, a patient's own (*isogenic*) immune lymphocytes are removed from their body, reprogrammed, grown in vast numbers, and reinfused back into the patient. The potential for cell therapies to cure cancers and other diseases is vastly increased by the availability of iPSCs and new methods, particularly CRISPR-based genetic engineering, to edit the genes of these cells. However, if the cells are used to repair or even make a brain, the locus of autonomy will be the brain and the person guided by the brain, rather than the individual cells used to repair or make the brain.

3

Persistence of memory

Via readily reversible *epigenetic* changes in DNA, protein, and RNA, somatic memories are passed from one generation of cells to the next. For reasons to be explained, epigenetic memory is only scantily transmitted across generations of people.

First, consider cellular memory. Epigenetic cellular memories are the mechanism by which a diversity of specialized cells differentiate from one fertilized egg, all equipped with largely the same starting set of genes. The fertilized egg first differentiates into three embryonic cell layers: ectoderm, mesoderm, and endoderm. The cells in these layers are progenitors of hundreds, or even thousands, of cell types, with the level of precision by which cell types are defined becoming more refined year by year. Neurons, epidermal cells, and melanocytes all remember their ectodermal ancestry. Although nearly identical in DNA sequence, their identities are epigenetically encoded as DNA modifications and as changes in proteins and other molecules. The epigenetic variations program which genes are expressed, the level of expression, and many other nuances of expression, including alternate forms of RNA transcripts and differences in the stability and translatability of transcripts into proteins.

The epigenetic changes retained in lineages of cells—from one generation of cells to daughter and granddaughter generations—include hundreds of thousands of reversible modifications of DNA nucleotides, for example the methylation and hydroxymethylation of cytosine nucleotides, one of the four DNA bases. The epigenetic changes also involve the proteins and RNA molecules associated with DNA and RNA that regulate the transcription of DNA into RNA, and that regulate RNA splicing, trafficking, and translation into protein. RNA epigenetics, also called epitranscriptomics, is the new frontier of epigenetics, the changes to RNA molecules enabling even more rapid and reversible plasticity of cellular function than modifications of DNA. Using long-read sequencing of unamplified DNA, it has recently become possible to read the methylation and hydroxymethylation patterns across any DNA region, one DNA molecule at a time. Via sequencing of DNA coprecipitated by antibodies directed against chromatin modifications (ChIP-Seq), the pattern of histone modifications or transcription factor binding in any region of DNA can be determined. All of this epigenetic biochemistry is mediated by other molecules that act as writers, readers, and erasers of the epigenetic code of DNA and RNA. Sequencing of RNA, including single-cell RNA, can then be used to correlate epigetic changes with actual effect on gene expression.

In the adult human body, epigenetic effects are pervasive and constantly in flux, for many purposes. For example, epigenetic changes cause more than one-half of genes to

Immortal. https://doi.org/10.1016/B978-0-323-85692-8.00003-4

21

cycle in expression in various cells of the human body, on a 24-h circadian cycle. Thirteen mutually repressing genes comprise the core of the molecular clock that controls the cyclical variations in expression of other genes (Bass and Lazar, 2016). One gene represses another in interlocking fashion like cogs in a watch, ultimately leading to their own repression, and calibrated to the 24-h cycle. Not surprisingly, the molecular clock built by natural selection, which Dawkins aptly called the "blind watchmaker," is phylogenetically ancient. In animals, the origin of the molecular clock appears to be monophyletic (meaning that it happened once). Core circadian clock genes in humans and fruitflies are orthologous (same gene, same function), reflecting our common ancestry, more than 800 million years ago and thus the circadian clock existed hundreds of millions of years before insects or mammals appeared on the earth.

Deep in the human brain is the suprachiasmatic nucleus, which is the body's master circadian clock. Together with external cues such as light, meals, activity, and melatonin release by Descarte's pineal, the suprachiasmatic nucleus tunes the circadian clock, keeping all the circadian clocks in cells throughout the body in phase with the light/dark cycle. If one feels badly off-kilter when jet-lagged, it is because all these little clocks, modulating expression of thousands of genes, and almost every aspect of the body's physiology, have become misaligned against the day/night cycle, making some travelers question whether they are zombies. However, because the clock mechanism is epigenetic, the effects of clock misalignment are reversible, unless a person carries a mutation disabling the clock mechanism, as some people do. When it works, adjustment of the circadian clock is an example of a rapid and highly advantageous epigenetic reset, recalibrating gene expression and physiology.

Due to its genetic and epigenetic programming, and in the environment in which it finds itself, the cancer cell "wants" to make more copies of itself. Like a protozoan in a pond, a cell that does a better job of making more copies of itself remorselessly outcompetes others, and every nuance of epigenetic or genetic change (mutation) that enhances its capacity is positively selected. The epigenetic and genetic modifications of a cancer cell may favor its growth over the growth of other cells in diverse ways, including unleashing the cell cycle of replication, tissue invasiveness, stimulation of vasculature feeding the tumor, or by disguising the cell from immune surveillance. Working together, these factors enable the cancer cell to outcompete cells. However, the same molecular modifications that enhance the tumorigenicity of cancer cells can be targeted in cancer therapies.

Second, consider the memory of the whole person compared to cellular memory. A defining difference between a human cell and a human is the framework of memory. People have memories of which they are aware (explicit memory) and memories of which they are unaware (implicit memory) but that may nevertheless shape their future behavior, sometimes more strongly than explicit memory. Memories are plastic, highly transmissible, and not always to be trusted. But they make us what we are. A trope of horror fiction is the zombie guided only by basic instincts and cravings. Fictional zombies look, and in some ways act, like people—much more so than real-life persons in a persistent vegetative state. But both are without self-awareness or memory, as far as is known.

A cell is a much simpler thing, but laboratory-grown multicellular organoids are beginning to bridge the gaps between cells, zombies, and people. The relevant question is what the cells are doing, not how many cells are doing it, or even whether they are put together in such a way as to appear human. The cell carries the DNA sequence of the person from whom the cell was made, and an epigenetic pattern that programs its function and responses. But, does memory of the life experiences of the donor persist in any meaningful way in a cell? In the end, the answer boils down to "no" but the complicated route to "no" requires tracing.

Cellular memory is always implicit rather than explicit. The cell knows not what it knows. Cellular memory can be plastic, and as has just been mentioned, this is cellular memory of the epigenetic type. More so than memories of which we are explicitly aware, cellular memory can be exceedingly resilient and long-lasting, and this is cellular memory of the genetic type. For example, in the form of phylogenetically ancient histone, ribosomal, and developmental switch genes whose sequences have been conserved for hundreds of millions of years, cells—and therefore humans—implicitly "remember" the way they were eons ago.

Cells thus have unconscious memories that are both short term and long term. They are unaware of what they remember, but within a human lifespan, somatic memories are held by cells and are passed from one generation of cells to the next. Cellular memory is the mechanism by which a diversity of differentiated, specialized cells differentiate from one fertilized egg, all equipped with largely the same starting set of genes. The fertilized egg first differentiates into three embryonic cell layers: ectoderm, mesoderm, and endoderm. These precursors further differentiate into hundreds, or even thousands, of cell types, depending on how one names them. Considering only cells of ectodermal origin, neurons of many types, epidermal cells, and melanocytes all have a cellular memory of their ectodermal ancestry and of later, more differentiated cells of which they may be daughters. Although all of these different types of cells are nearly identical in DNA sequence, their identities as a cell of one type or another are encoded as DNA modifications and as changes in proteins and other molecules. The *epigenetic* changes determine which genes are expressed, the level of expression, and many other nuances of molecular expression, including alternate forms of RNA transcripts and differences in the stability and translatability of transcripts into proteins. These changes in DNA expression are what make one type of cell different from another.

Whereas genetic memory is long term and transgenerational, epigenetic memory usually is short term and mainly relevant within a somatic generation. The biology of pluripotent cells—cells capable of differentiating into any other cell—have taught us much about why epigenetic memory is almost entirely erased between generations of people. Whole persons begin with one cell, a fertilized egg. To artificially make a pluripotent stem cell, it is necessary to reverse many of the epigenetic changes that make a differentiated cell—in effect to clear cellular memory of all the epigenetic changes that have been made during the history of its somatic lineage. This is not as easy as resetting the circadian clock, but the body erases most epigenetic memory early in development.

Early in embryogenesis most of the cellular memories of our parents, and all of the implicit and explicit memories encoded in our parents' brains, are wiped clean. In this sense, but not the main sense in which it was intended, the blank slate hypothesis promulgated by some psychologists was accurate. Without cellular *tabula rasa* the embryo would not be able to start over to generate the three primordial cell layers, and the myriad cell types that differentiate from these cells. Due to the erasure of most epigenetic information, succeeding generations take delivery of the consequences of genetic mutations but benefit or endure little from what is epigenetically encoded.

Behavioral *tabula rasa* due to epigenetic erasure is generally advantageous, but more importantly there would be no physical mechanism by which complex behaviors could be encoded within a single sperm or egg and faithfully transmitted to the new brain. The behavior encoded in complex neuronal ensembles would have to be translated into molecular language within the germ cell and then back-translated to create a neuronal network in the developing brain encoding the same behavior. What could possibly go wrong? A famous anecdote revealing the frailty of translation, even from one spoken language to another, begins with a meeting between apocryphal Japanese and American trade delegations. Trying to apologize for an oversight, the American used the phrase "Out of sight, out of mind," which was promptly translated "Invisible idiot." This example, with its missing comma, also shows how easily meaning can be corrupted or reversed by a mere error of punctuation. Imprecise translation is not merely useless, it is dangerous: "I ate, Mother."

In the brain and through development from helpless altricial infant with a few rudimentary behaviors to adults capable of multifarious, intricate ones, translation from the molecular language of genes and epigenes to behavior is accomplished stepwise via the growth, distribution, and networking of constellations of neurons. It takes time, and more often than not the neuronal networks are shaped by practice and reiteration, only gradually optimizing the behavior or skill. As Jane Austen's Lady Catherine de Bourgh declared, "If I had ever learnt, I should have been a great proficient." On her propensity for music, and as might be influenced by genes and epigenes, "There are few people in England, I suppose, who have more true enjoyment of music than myself, or a better natural taste." However, the missing key to musicianship for Lady Catherine and so many of the rest of us, is …practice: "I often tell young ladies, that no excellence in music is to be acquired, without constant practice. I have told Miss Bennet several times, that she will never play really well, unless she practices more…".

It is instructive that genes that influence behavior do so via networks of causality, and usually enhance fitness by altering propensities that make a behavior somewhat more or less likely. Specific behaviors that are genetically encoded—such as a rat's fear of snakes, or mating behaviors—are few in number and have been honed by many generations of natural selection that have shaped ensembles of genes. There is no specific gene for piano playing because the pianoforte was invented only in the 19th century. Translating, it is highly unlikely that a person can molecularly encode a complex behavior in such a way that the code can be translated by a sperm or egg, and then back-translated. On the other hand, a propensity for music could be ancient.

Each newborn begins life without the full burden of memory and experience of their parents. Their parent might have suffered burns, but the child will not fear fire unless taught or directly experiencing a burn. At the present time there are enthusiastic reports claiming that cross-generational transmission of complex behavior can be epigenetically encoded. For example, Dias and Ressler reported that conditioned place preference of a male rat could be transmitted to rat pups (Dias et al.). We can think of this as behavioral Lamarckianism, referring to the French naturalist Jean-Baptiste Lamarck (1744–1829), who was an early advocate of evolution, and a predecessor of and acknowledged influence on Darwin. Lamarck fought Cuvier's dogma of species stasis—the amply refuted idea that new species do not evolve. Unfortunately, Lamarck also introduced the more tenacious idea that animals' experiences and uses of body parts leads to changes transmitted to the next generation. In succeeding generations, giraffes' necks and tongues elongated because their ancestors had stretched to reach leaves. Similarly, we might expect that after generations of running mazes, rats would become good at it, from birth.

The second wave of Lamarckianism was in the Soviet Union. A generation of Soviet genetics was hamstrung by imposition of Lamarckian doctrine as reformulated by Lysenko. Perhaps 21st century transgenerational epigenetics will succeed where Lamarck and Lysenko failed. However, although the new studies showing transgenerational transmission of behavior are intriguing, there is no plausible mechanism by which a parent's training can be transmitted to sperm or egg, survive the systematic erasure of epigenetic programming early in development, and then shape the development of the brain towards a specific behavior such as conditioned place preference.

History does not repeat itself, but rhymes and mimes: Cultural transmission and adaptation have been compared to Lamarckian evolution. However, the story of transgenerational epigenetic transmission is perhaps a better example of the frame of reference problem in which the minutiae of one generation are forgotten by the next. Cultural icons are forgotten and replaced by new ones. The generations can lack a common language of communication. A standard of writing, which most authors fail at in one way or another, is to eliminate references that are likely to be meaningless to subsequent generations.

In science, the transgenerational transmission of knowledge is particularly difficult, because so many discoveries of the moment are either discovered to be wrong, or are superseded. Whereas Karl Popper taught that scientists attempt to falsify their hypotheses, Kuhn theorized that science proceeds via revolutions, old dogmas eventually being overwhelmed by new paradigms. Many of the most important expressions of science are, like Stravinsky's *The Rite of Spring*, violently rejected when they are first heard. Discoveries in this category are legion: Wegener's continental drift, the transposons of Barbara McClintock, the "dinosaur heresies" Robert Bakker and John Ostrom that these ancient creatures were warm-blooded and active, and later the insight by the father-son team of Walter and Luis Alvarez that those same dinosaurs were wiped out by a meteor, and many more. Wegener's theory was met with derision. On the other hand, when Albert Einstein published five epochal papers in 1905, culminating in the $E=mc^2$ relationship that unified matter and energy, the response of the physics community was basically silence, until one leading figure in the field, Max Planck, recognized the virtues of Einstein's discoveries and advocated them to the wider community. Although it is most accurate to say that modern science (after the classical era) is governed by the iron rule of empirical data, some of Einstein's theories, at least initially, and many of the most influential theories of physics (e.g., string theory) are influential because of their logic and beauty, the empirical support coming later. By way of contrast, biology has its important theoretical disputes (for example, the extent to which Motoo Kimura's neutral theory explains population genetic variation and molecular evolution). However, biology is to a large extent data driven. Einstein's discoveries, like many other complex findings, became simplified into memes such as $E = mc^2$ that have percolated through the discourse of the world community and influenced it profoundly, even though the people who repeat the meme have only a dim understanding of it. However, they have some understanding; they might know that it has something to do with matter, energy, and the speed of light, even if they do not understand that it was Einstein's discovery that it was the speed of light that was constant, and that mass becomes enormous, and time slows, as one approaches the speed of life. In this regard, people are very different from cells, and also the same. Cells of a certain kind have a common chemical language and a chemical language for communicating with cells of other specific types.

Within a society, there are also common icons and languages of communication. As was attributed to Charles V, Emperor of the Holy Roman Empire, "I speak Spanish to God, Italian to women, French to men, and German to my horse." who had said something similar. Because at birth humans do not have heroes, cultural icons, or even languages, they are—from the perspective of society—more pliant and programmable than cells. People can be programmed and deprogrammed, quotes can be changed as needed, and the transitions are most facile and dramatic across generations. This is not how cell communication works, but it is how human communication works. After Lamarck's meme of transgenerational epigenetics was inserted into the human consciousness, it was inevitable that it would resurface. When it did, it derailed Soviet genetics. Time, and data, will tell whether present-day neuroscientists and genomicists trying to advance transgenerational epigenetics will succeed where Lysenko failed, and doubters such as myself shown to be obstacles to progress!

Once a culture of growing cells, be it HeLa or iPSCs or some other cell used for a cell therapy, is initiated, it is usual that some of the cells are cryopreserved for later use or as insurance against technical mishap. These cells can be cryopreserved for years and probably centuries, later to be thawed and grown with full vitality and used to make whole multicellular organs and persons. Technically this is highly feasible, but, whether fortunately or unfortunately, the memories and experiences of the donor are lost. The cells, or the organoids or persons made from them, will not be the donor person, not having the epigenetic memories of the person from whom they were taken. By comparison, the whole body or brain does have those memories, if we could preserve it. However, it is far more difficult to cryopreserve a whole body or brain that would enable more than a false resurrection. The ancient Egyptians and others tried, mummifying bodies and carefully removing organs and placing them in canopic vessels. Right up to the present day, millions of bodies are preserved and sealed in watertight caskets prior to burial, but anyone investing in cryopreserving their body after death is aware that these procedures will not enable resurrection, in the way Lazarus was resurrected. In nature some animals, but mainly miniscule ones such as the tardigrade (water bear), have mastered the Lazarus trick. Unfortunately, a frozen human body or human head might be useful for various purposes, including making a DNA sequence-based copy of the original. Unfortunately, it is unlikely that human heads cryopreserved with methods presently available could be used to bring the person back to life or bring back to life the donor person. Ice crystals formed during the freezing process hopelessly disrupt neurons and their connections. Paradoxically, the body's single cells are at once extremely fragile, but when frozen or grown in the laboratory under special conditions, potentially immortal. But memories embedded in neural nets are fragile. If one is obsessed with immortality, the best advice is to hope that some religious faith is correct (whether about afterlife, resurrection, or reincarnation), create a work that lives after you (a pyramid, a book), make children (and grandchildren), make friends and influence people, be a hero (like Achilles), or transmit an idea that takes hold.

Could memories and selves be transmitted from one body to the next?

The fabled homunculus was a miniature person in the forecastle of the mind, steering the ship. A homunculus could also grow into a whole new person, explaining procreation. In the 21st century the observation that there is no homunculus, cell, or organ within any of us might seem obvious. However, as shown by how people talk about mind and self, belief in the homunculus lives implicitly, even if the name has fallen out of explicit use.

Building the homunculus? In biology, only the unwise ever say never. Humans might one day soon find a way to install in the brain a cellular or cybernetic device controlling the brain in ways speculated by science fiction writers. The potential for control by cybernetic devices is obvious—both miniaturization and the ability to electromagnetically transmit commands to microelectronic devices would enable the types of control feared by many paranoid schizophrenics. More subtle is the option of dispensing with the electronics and installing multicellular command and control biocomputers. The synthetic cells that would comprise such devices would already be well suited to receive information and issue instructions to native cells and talk to each other. Organoids are grown in cells-on-chip models. Most of the genetic engineering necessary to turn synthetic cells into artificial biocomputers has already been invented and successfully prototyped, as described by Joanne Ho and Matthew Bennett, bioengineers at Rice University (Science, 2018). These cogs include logic gates, timers, counters, and memory devices that can encode events—and as has already been done, bits and bobs of a movie—directly into the DNA of "lowly" bacteria. Bioengineered cellular parts are already capable of performing complex calculations, raising the prospect for real-time surveillance and consequent decision-making and command. Whether a bioengineered cell, interconnected group of cells, or organoid would be regarded as a self is questionable. This would depend on its capacities and perhaps its ability to effect change in the external world.

Whereas artificial biocomputers from single cells are a goal that may only be realized in 10–20 years, evolution accomplished this trick hundreds of millions of years ago and has optimized it ever since. As will be seen later, things sometimes go wrong, but the unfolding of development of a complex organism from a single cell is, at this point in our understanding, close to miraculous. Humans and indeed all metazoan life begin as single cells without the slightest resemblance to the adult organism in form or capacity. The single cells made by fertilization of an egg do not have the ability to think on their own in the way their multicellular adult forms may, or even a remote chance for long-term survival. Through cell proliferation, differentiation, and amazing transformations involving masses of cells: blastocyst, gastrulation, neural crest formation, and segmentation, one cell becomes a multicellular organism with organs, a brain, and capacity for thought.

When cultured cells such as iPSCs (induced pluripotent stem cells) or neurons derived from them are used to repair, or even make, a brain, the locus of moral autonomy will be the brain and the person guided by the brain, rather than the individual cells, once cultured in a flask. Belief in free will does not necessarily follow from brain/mind dualism. The homunculus or spirit sitting at the controls of the brain directs its actions, and in effect the cells and circuits of the brain comprise the sections of an orchestra under the conductor's baton. However, what animates and directs the homunculus? In the early days of neuroscience the homunculus model was only slightly complexified, and while the foundations of behavioral causality were introduced, mind/brain dualism was not abrogated. Behavioral determinism was fully compatible with mind/brain dualism.

Freud, who in superego fashion vigilantly watched his disciples for deviance, introduced the id, the ego, and the superego, a heuristic model that continues to live on in social science and literature even though it was discarded long ago by neuroscientists. Freud's ego may be regarded as the conscious, controlling homunculus that balances desire (id) and the moral constraints with which we are inculcated (superego). Freud's model of the mind remains influential but was superseded by the isolation of the cells, neurocircuitry and neurotransmitters responsible for executive cognition and impulse control, emotional salience and emotional valence, and memory, and as well the neurobiology of more "mundane" functions such as sleep, sensory, appetite, thirst, respiration, and movement.

Freud would have been fascinated to know that depressive thoughts can be triggered not by the tragedy of childhood trauma or defective relationships with one's mother or father, but by depletion of the neurotransmitters dopamine and serotonin, as was first discovered some 40 years ago as a side effect of antihypertensive drugs. Also, the veil of depression can—in some patients—be lifted by electric stimulation of a specific region of the frontal cortex, as discovered by Helen Mayberg, Sid Kennedy, and others, about a century after Freud. Depressive thoughts lead to changes in neurotransmitter and endocrine function, leading to self-perpetuating, vicious cycles (not "circles"), from thoughts to effects on neurotransmitters and brain function and then back again. We are our brains, our brains are what we remake them, and then our brains in turn remake us. When the origin of depressive thoughts has been shown to be a neurochemical or a neural pathway, and reciprocal relationships between brain and cognition are understood, it becomes more difficult to adhere to mind-brain dualism—the brain is the mind.

Where Memory and Self Reside: Freud would have been surprised to understand the reciprocal interplay between consciousness and recollection of emotional memories and the neurocircuitry of the amygdala, hippocampus, and more distributed parts of the brain in which memories reside and are given emotional valence. There is probably no challenge that is more complex, but in the past decades Joseph LeDoux, Elizabeth Phelps, and Daniela Schiller at New York University and others have tracked down key components of the neural networks that are responsible, and discovered recipes for the manipulation of emotional memory that would been unavailable to Freud and his contemporaries, limited as they were by the knowledge and tools of their era.

The molecular basis of consolidation of memory was uncovered by Nobel laureate Eric Kandel, who studied the ability of the motor neuron of the sea slug (*Aplysia californicus*) to learn to trigger withdrawal of its siphon. The same biochemical events happen in cells of the human brain to enable us to learn. Without getting into unnecessary detail, these events include specific types of intracellular signaling. Memory also requires longer-term changes in receptor expression on the membranes of neurons, protein synthesis, and the strengthening or weakening of synapses by the growth and involution of parts of the neuron, as if the fingers and thumbs had gotten larger or smaller to deal with changing demands in the way one was using one's hands.

It is facile to say that a cell is not a person because it does not carry self or memory, because memory is fragile. Amnestic individuals forget their selves, and what has previously motivated them, and under twilight anesthesia, when they are temporarily unable to encode new episodic memories, become highly suggestible, as if they do not have free will. Most memories that are registered by our brains are quickly lost, and their consolidation is easily disrupted by the smallest distraction. Of course this is invaluable because otherwise our brains become packed with useless information. How many "Smiths" are in the phone book, and what are their given names, addresses, and phone numbers? We are only beginning to understand the nature of human consciousness and memory, but have learned that it is in many ways fragmentary and fleeting.

The role of cellular memory in development At the beginning of the lives of most multicellular organisms there is a pluripotent cell—for example, a fertilized egg—and some unthinking molecules directing the action. The developmental transformations the progenitor cell and its descendent cells undergo are phylogenetically extremely ancient, being shared across life-forms whose common ancestors lived hundreds of millions of years ago. Also, to some extent these embryonic changes recapitulate evolutionary changes necessary for the self-assembly of new life-forms, connecting humans to the simpler life-forms from which we evolved, and telling us that if we rewind history a billion years to the first multicellular life, there was also never a homunculus at that point. A miniature version of a person, or a fish, existed nowhere in this matrix of molecular and cellular causation—the person, or fish, only emerged at the level of the interaction of these components. But to make development happen, cellular memory is needed. Genotype is shared by all cells of the body, more or less, but cells vary wildly in the expression of these genes, enabling their specialization. Therefore developmental biologist Conrad Waddington proposed that long-lasting genetic memory is supplemented by more plastic epigenetic memory, enabling cells to differentiate into specialized types and to be reprogrammed for new forms and functions.

Some genetic and epigenetic switches have more profound impact than others, and within the body there are some genes, and cells carry them, that are more sensitive to these changes. A proto-oncogene is a gene that can readily be mutated or reprogrammed to cause cancer. Because of their functions, proto-oncogenes are "ready" to be a cancer gene, should they be repressed, unleashed, or modified. Genetic mutations involving specific DNA nucleotide substitutions, smaller and larger insertion/deletions, translocations, or loss or gain of entire chromosomes or large pieces of chromosomes can all set free a proto-oncogene. A single nucleotide substitution or small insertion/deletion can easily abort the translation of an RNA into protein, convert the protein into a less active or super-active form, or damage a site suppressing or enhancing expression of the gene, or the processing and transcription of the RNA made from the gene. Certain genes, such as the *P53* gene, are recurrently affected in particular cancers, and as constitutes a vulnerability, with different mutations in the same gene in different patients, but also a starting point for treatments targeted against cancers with mutations in particular genes. Is the gene itself a treatment target? Yes, but reversal of such small genetic changes would require specific genetic surgery, which on a natural basis occurs so rarely as to deserve the adjective "never." More often it is a gene product that is targeted, and for example by an immuno-therapy against a protein not normally expressed, or by intervention in a biochemical pathway, sometimes in an immune cell monitoring cancer cells or checking the action of other immune cells.

Translocations placing genes in abnormal locations where they are switched on by powerful promoters are common in cancer, especially in cancers of immune cells in which the natural translocations necessary to produce immunoglobulin diversity can go awry, leading to a series of highly characteristic translocation mutations in these cancers. Many translocations have been associated with cancers, but prototypes would include several translocating regions of chromosomes harboring proto-oncogenes to chromosome 14, where they are massively transcribed under the control of an immunoglobulin promoter. The immune cell whose molecular machinery has been evolutionarily honed to make large amounts of immunoglobulin instead makes large amounts of the oncogene. Examples of chromosome 14 translocations leading to specific cancers of the immune system are chromosome t(8;14) *CMYC*→ Burkitt's lymphoma, t(11;14) *Cyclin D1*→ Mantle cell lymphoma, and t(14;18) *BCL-2*→ Follicular lymphoma, the latter being responsible for 90% of cases.

In addition to the largely irreversible point mutations and small deletions, the translocations and larger deletions that irretrievably delete normal copies of genes are even more difficult to reverse, or irreversible by natural processes. However, the epigenetic modifications that may trigger cancer or be mechanistic in the action of the genetic changes are surprisingly easy to reverse. Epigenetic reversal is the basis of many new cancer therapies in which the expression of key genes is repressed or modified, or in which the cancer cell can be tricked into differentiating into more mature, benign cells, as was first accomplished almost four decades ago (Rovera et al., *Science*, 1979) for chronic myelocytic leukemia and is being applied to many types of cancer. Other natural defenses, such as killer T cells, and medical approaches, such as surgery, chemotherapy, radiotherapy, and immunotherapy, kill the cells. All these measures defeat the cancer cell only in the narrow sense that it is unable to outcompete and overwhelm other cells of the body. However, in a broader sense the rehabilitation, or even execution, of an unrepentant cancer cell usually makes it more likely that the cell's own genes are transmitted to succeeding generations.

Fortunately for the transgenerational well-being of our species, the genetic and epigenetic changes that would cause cancers and other problems are almost entirely kept under control, if not eliminated. A few new (de novo) germline mutations are transmitted each generation, but the deleterious mutations are rare enough to be dealt with by natural selection, which expunges their transmission.

Natural selection could not possibly remove the millions of epigenetic changes, so a mechanism exists to rid the germline of the vast majority of epigenetic changes to DNA, RNA, and proteins. Rapidly accumulating epigenetic changes that would cause cancer and many other types of problems are almost completely erased and expunged early in embryonic development. *Tabula rasa*? Not quite, but even to induce a pluripotent stem cell for therapeutics or research it is necessary to reverse many of the epigenetic changes that make a differentiated cell—in effect to clear cellular memory. This is not as easy as resetting the circadian clock, but the body erases most epigenetic memory early in development. Early in embryogenesis most of the cellular memories of parents, and all of the implicit and explicit memories encoded in the parents' brains, are wiped clean. In this sense, but not the main sense in which it was intended, the blank slate hypothesis believed in by psychologists who discounted the role of inheritance in behavior was accurate. Without epigenetic cellular *tabula rasa*, the fertilized egg would not be able to start over to generate the three primordial cell layers and the myriad cell types that differentiate further down the developmental path. Despite the erasure of most epigenetic memory, succeeding generations still have to endure, or benefit from, the consequences of genetic mutations, but they benefit or suffer from little that was epigenetically encoded.

Behavioral *tabula rasa* is an advantage, but whether or not beneficial, it happens. Under natural conditions there is no physical mechanism by which complex behaviors could be encoded within a single sperm or egg and telepathically communicated to the future brain. Each newborn person begins life without the full burden of memory and experience of their parents. Their parent might have been burned by fire, but they will not fear fire unless taught or unless they directly experience it. As mentioned a little earlier, at the present time there is a host of enthusiastic scientific reports claiming that epigenetic memories can be transmitted intergenerationally. For example, a study by Dias and Ressler found that conditioned place preference of a male rat could be transmitted to offspring (Dias et al.), via the sperm. This would require the sperm cell to be imprinted (presumably via a highly specific molecular message in the blood of the father) with its own little memory of a complex learned behavior of the father, and then to somehow transmit this information to guide the brain to develop the specific circuitry to avoid the "bad" side of the chamber.

We can think of imprinting of past life onto brain and body as behavioral Lamarckianism, in remembrance of French biologist Lamarck who believed that animals' experiences and uses of specific body parts led to changes transmitted to the next generation. In succeeding generations, the giraffe's neck and tongue would elongate if ancestors had stretched to reach leaves. A generation of Soviet genetics was crippled by imposition of Lamarckian doctrine as reformulated by Lysenko.

Perhaps 21st century transgenerational epigenetics will succeed where Lamarck and Lysenko failed. However, although the new studies showing transgenerational transmission of behavior are intriguing, there is no mechanism by which the sperm can be made aware of what the brain has learned, much less transmit this information to neurons of the newborn rat such that later in life the rat would exhibit the specific learned behavior.

Reincarnation, past lives, and all that comes with it: The implausible feat of memory transmitted across the generations proposed by some scientists was echoed, in a charming way, by actress Shirley MacLaine. *Above the Line: My Wild Oats Adventure*, which I am pleased to advertise and to further remember MacLaine herself in this book. MacLaine displayed the remarkable ability to recall memories from some two million years ago, when she was a resident of ancient Atlantis. Unfortunately, and as should be a lesson to all of us, the "star nations" who built Atlantis were too greedy, and were duly punished by a series of deluges that sank the island continent. Over long epochs, many interesting things happen. During pre-Atlantan "Lemurian" times, MacLaine was an androgynous being, with both male and female genitalia but procreating through the power of the soul. This state of affairs ended when MacLaine and other Lemurian troublemakers decided to separate yin from yang, male from female. It cannot be ruled out that MacLaine remembered past lives better than most because her lives were more interesting, but her memories did seem to have come in handy. When MacLaine, in her present incarnation, was in a romantic relationship with Swedish Prime Minister Olof Palme, it occurred to MacLaine that the whole affair felt so familiar because she, poor Moorish peasant with a knack with impotent men, also had sex with Palme 1200 years before. However, at that time, Olof Palme was Charlemagne, Emperor of the Holy Roman Empire. In other past lives MacLaine was a harem girl for a Turkish Pasha, a gypsy living with Coptic Christians in the Spanish hills, a medieval warrior, an orphan raised by elephants, a Japanese geisha, and a model for the Postimpressionist painter Toulouse-Lautrec.

If one accepts the theory of reincarnation, MacLaine's memory of a dozen past lives is hardly excessive. Indeed, it is minimalistic. It does not seem strange that MacLaine remembers so much since Lemurian times, but rather it is disturbing she remembers so little. Human generation time has always been a bit controversial, the idea being that generation time has lengthened in modern times, but by analyzing Neanderthal gene flow produced by admixture with *Homo sapiens* some 50,000 years ago, with subsequent recombination of the introgressed Neanderthal DNA segments with the rest of our genomes, David Reich and colleagues were able to determine that over the past 45,000 years the average generation time has been 26–30 years. Using this generation time, in only the past 45,000 years, any person would perhaps have been reincarnated some 1500 times, and Shirley MacLaine, or anyone, would have many more memories over two million years, which MacLaine's memories go back to. Of course, there might be even more memories of past lives spent as other animals, there not having been enough human ancestors to have carried the spirits of all the humans now living on our crowded planet, and the generation times of other animals generally being shorter. Or perhaps the soul from time to time takes a time out from reincarnating.

To understand how reincarnation would work in practice, it is also useful to consider its leading examples. The first Dalai Lama was born in 1391, and since then there have been 14. These numbers suggest that, at least for Dalai Lamas, reincarnation time is slightly longer than generation time. At the age of 2, little Lhamo Dhondup was recognized as the latest Dalai Lama on the basis of his demeanor: for example, he would beat quarrelsome people with a stick. As further proof of divinity, the boy successfully identified people and possessions from the past lives of Lamas.

When Lhamo Dhondup and 15 other candidates, all boys, were taken to Lhasa to meet the governor, Lhamo alone did not cry and cling to his parents. He took the only vacant seat and while other boys wolfed down treats, took only one snack for his great-uncle. A clinching proof was that he identified the governor (from memoir of Diki Tsering in *Dalai Lama: My Son*, Compass Books, 2000). Nowadays, the Dalai Lama is reticent to speak of memories of his past lives, wisely saying that he has trouble remembering what happened yesterday.

Despite the Dalai Lama's humility, as a way of transmitting memories across generations, reincarnation is probably more likely than epigenetic transmission of memory down the generations from ancient ancestors. The crushing number of epigenetic transmissions helps explain why. Whereas reincarnation involves one ancestor per generation at most, epigenetics would inflict the burdens of ancestral memory of all progenitors of preceding generations. Anyone reincarnated since the admixture with Neanderthals 50,000 years ago might have to sort out the burden of 1000 to 2000 memories. However, epigenetic transmission of memory would be worse. Any person has had two parents, four grandparents, eight great grandparents, and with no "time out" as is at least theoretically possible with reincarnation. Resurrections might be interrupted, but epigenetic transmission occurs generation by generation (unless it does not happen!). Discounting the effects of inbreeding (consanguinity), and as is a reasonable approximation until the effective breeding size of the human population, which happens to be approximately 10,000, is approached, the number of ancestors increases exponentially each generation. At generation 13 and thereafter 8000–10,000 ancestors per generation would have contributed to our epigenetic ancestral memories, if indeed we ever had them.

Although Shirley MacLaine and many others are pleased to remember their past lives, some religions have recognized that reincarnation and ancient memories are burdensome. And, they are doing something about it. Buddhism attempts to break the cycle of reincarnation permanently. A modern-day religion with millions of adherents worldwide,

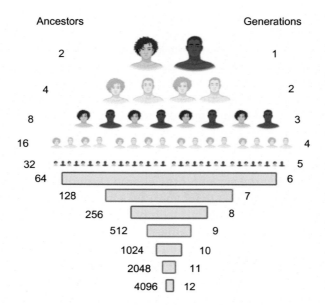

Ancestors		Generations
2		1
4		2
8		3
16		4
32		5
64		6
128		7
256		8
512		9
1024		10
2048		11
4096		12

Considering only the past 12 generations, each person has had some 8000 ancestors. Past generation 12, any person had approximately 8000 ancestors per generation, that number being limited to the effective population size of human populations. The implication is that our genomes carry small quotients of genetic memories of any ancient ancestor. Our genomes could carry only small quotients of epigenetic memory, if they carry them at all, any individual ancestor's epigenetic memories having been diluted by the epigenetic memories of other ancestors. *Created with BioRender.com.*

the Church of Scientology, founded by science fiction writer L. Ron Hubbard, recognizes that ancient memories are bad. As the Church of Scientology likes to point out, the vast majority of the world's population believes in reincarnation, but only Scientology (all information from the official website https://www.scientology.org/faq/scientology-beliefs/reincarnation.html) is expunging ancestral memories for people who will or cannot stop reincarnating. "Scientology assists its followers to rid themselves of ancient engrams, thus giving them tools to handle the upsets and aberrations from past lives that otherwise are bound to adversely affect the individual in the present."

Where consciousness, if not personhood, resides: It is unlikely that frozen cells or even a frozen head could be used to restore a person. Humans are not tardigrades. If one is obsessed with immortality, the best advice is to hope that some religious faith is correct (whether afterlife, resurrection, or reincarnation), create a work that lives after you (e.g., a pyramid), make children (and grandchildren), make friends and influence people, be a hero (e.g., Achilles, Lenny Skutnick), or transmit an idea that takes hold. However, cells can be cryopreserved for years and probably centuries, and then thawed and grown with full vitality. Paradoxically, the body's single cells are at once extremely fragile, but when frozen or grown in the laboratory under special conditions, potentially immortal.

Summarizing what we know about the potential of any cell, even if we discount reincarnation, vital force, and transgenerational epigenetics, we should be cautious in deriding the information contained within one cell. This genetic and epigenetic information is uninformative for many aspects of the whole person, but it is specific to the individual, and—in the epigenetic and genetic information the cell contains—can be a legacy of that person or a tool to help rebuild that person.

4

Hel: Goddess of death and life

Remember tonight… for it is the beginning of always
Dante Alighieri

Among the Norse gods, startling physical transformations occurred across and within the span of one generation and some families were multifarious both in physique and temperament. As spoken in Norse oral tradition as written in the Edda, Hel is a goddess of death and life. She was daughter of the giantess Angrboda and Loki, the trickster, and her siblings were Fenrir, the archetype of wolves, and Jormungang, the world serpent. By Sigyn, Loki also fathered Narfi and Vali, the latter of whom was transformed into a wolf. Loki, in the form of a mare, also gave birth to the eight-legged horse Sleipnir, the father in this case having been a stallion. As implied in Norse mythology, Hel had the transforming power to bring the dead to life (although, deceived by her father, Hel did not use her powers to resurrect Baldur, a much-loved god killed by her father). Nowadays, genetic engineers and biomechanics of other sorts are engaged in similar transformations.

Nowadays, genetic engineers and biomechanics of other sorts can perform miracles of the type Hel was said to be capable. As tools for these transformations, and as sources of information to design new life-forms, modern bioengineers take their portion of organs, skeletons, DNA, and cells. To use them better, some cells, such as the cancer cells used to make HeLa, are immortalized. Like the goddess Hel, but eschewing magic for technologies such as cell therapy and chemo- and mechanotherapies (organoids, artificial organs, nano-technologies, and suchlike), modern-day scientists would stave off death and even undo it via genetic and cybernetic resurrections. Even as man-made mass extinction of natural species proceeds in the Anthropocene, scientists use cells and genetic tools to make chimeras and life-forms such as the world has never seen.

People are absorbed with the mysteries of birth, maturation, relentless decay, and inevitable death—the four seasons of life being likened to the passing of the year. We usually believe people are fortunate to experience them all, and to witness the cycle of renewal in the next generations. In keeping with the cyclicity of life and inevitability of death, a surprising number of people claim they would not want to live forever. It can be said that a person begins as a single cell and ends as trillions dying. But is that true? Usually, people behave as if they fear death.

A person is a machine constructed of molecules, cells, and organs. Under the right conditions, cells can be preserved and replicate essentially indefinitely. What part of a person continues if their cells go on living? Is Henrietta Lacks in some sense still alive? If a heart is transplanted into a recipient, does the donor continue? What if the brain is transplanted?

Immortal. https://doi.org/10.1016/B978-0-323-85692-8.00004-6

What about a part of the brain? It is one thing, and probably a measure of mental acuity and health, to accept something that is inevitable, but what are the progeny of our imaginations if death is not inevitable, or if by some trick of the mind we can deny it? Naïve reductionism teaches that a body is a sum of parts, but nuanced reductionism instructs that a manifest of the parts and a description of their nature are key to decrypting the workings of the whole and delimiting its boundaries and nature.

Some mid-20th century physicists gave the nature of life deep thought and helped found the discipline of molecular genetics. Speaking partly in jest, what did *they* know? Experimental physicists are more likely to put creatures (after a long life and peaceful death) into high-speed accelerators, smash them against targets or one against another, and analyze velocity, trajectory, mass, and composition of the fragments. But maybe they did know something. Increasingly, biologists use techniques both identical and analogous but more apt for the study of so-called "living systems." Biologists use physical methods such as sonication, dispersion with detergents, digestion with enzymes, laser-guided cell sorting, and dissection with scalpel and forceps to isolate organs, tissues, cells, and molecules. They expose cells and molecules to energies from all parts of the electromagnetic spectrum to peer into living things and trigger responses: for example, they can genetically engineer photon-responsive channels into a cell's membrane so that the cell can be activated or inactivated at will. The study of the constituents of so-called life is incredibly complex and detailed as compared to physics, and as a discipline biology demands different skills and mindsets. Chemistry is complex, but biochemistry is complexity cubed. Each genome, organism, organ, cell, cell organelle, and biomolecule is a world, and increasingly these worlds, and our appreciation of gaps in our knowledge, are coming into sharper focus.

So, what is life? Humans are a culmination of interactions of molecules and cells, or most meaningfully understood as expressions of these interactions that can increasingly be measured, modeled, and manipulated. The genome is a 500 million year old toolkit of genes encoding development. In the past several decades, genes that trigger the formation of limbs, eyes, brain, and other organs have been identified. The stepwise activations and deactivations of these genetic switches in the right places and times create the morphological diversity of life—what Darwin called "endless forms most beautiful." Increasingly, embryology is studied in artificial cellular, organoid, and whole animal models. As told by Sean Carroll in *Endless Forms Most Beautiful*, a butterfly's spots, an animal's limbs, and eyes, and an insect's wings, segments, antenna, and eyes can all be made to appear and disappear via manipulation of single genes.

All life on earth probably evolved from one cell—and ultimately one self-replicating molecule. In the beginning was a cell, and in every generation a whole person is made from a single cell. The persons made from those single cells may be more ephemeral than we like to think, despite our evolutionarily favored inner sureness of continuity. Our brains perform a biological reencoding of self every day of our lives—transmitting continuity of self from day to day despite the death and birth of neurons and constant neuroplastic changes in dendritic spines, synapses, and molecules. What if the self could be reencoded, or copied, or mimicked, outside the brain, as depicted in science fiction? Increasingly, patterns of thinking and memories can be extracted, and neuroscientists have succeeded in encoding memories (admittedly, crude memories) without an animal having experienced the event in real life.

People often have false memories, which is one challenge in detecting liars. Technically speaking, most of the dozens of children who reported their false memories of abuse in the well-known McMartin case were telling the truth, as far as they knew it. Increasingly, memory and self can be altered without inserting electrodes into the brain. Virtual reality, unscrupulous "therapists" and cult leaders, and one's own dreams are sufficient to remake the self. In the near future, and in matter to energy and back fashion, memories, personality, and sense of self may be efficiently reencoded into biological and electromechanical vessels, to the point that—in Turing test-like fashion—the copy cannot be distinguished from the original, and the copy will have "self-knowledge" and self-certainty that it is the original. Such procedures are at this point science fiction, or extrapolation, but the fragile nature of the self and the real potential that it can be partly reencoded point to the weakness of the boundary between self and cells, molecules, and the information they encode.

The cell and molecule-diminishing argument that a cell or molecule is only morally significant in the context of a whole person can be extended to any "whole person." Later, we will dismantle that argument. For now, note that any person on their own, Robinson Crusoe before Friday, can scarcely survive. In survival scenarios, and as anyone knows who has watched a survival reality show, few people, even if specially trained and in peak physical condition with all immunizations up to date, last more than a few days before succumbing to lack of food, water, fire, or shelter, or suffering a crippling accident or illness. Even if they survive, in any historical or evolutionary sense, the person alone on an island has no significance, unless their message in a bottle is later read. Unobserved and unrecorded, the isolated person is as indefinite as Schrödinger's cat snug in her box. But people are not islands or alone on islands. They comprise a superorganism whose parts are interconnected by language, culture, and technology as much as by our genomes. The human superorganism constantly sheds, renews, and replaces pieces of itself. As proven by its self-destructiveness, it is only partly self-aware, but is evolving into something with knowledge and awareness beyond that of itself, any person who is a constituent, or any artificial intelligence that may be created by it. Even if our moral significance is diminished by how we are used by this larger thing, we also have moral significance only in relation to the evolving and continuing whole of humankind.

People are constantly imputing moral significance to derived pieces of ourselves, and this is a subtext of moral conversations, with growing implications for how we treat people, tissues, cells, DNA, and information decoded from the living. Do these pieces of selves have moral standing, and if so under what circumstances? What does it mean to be a self or a personally identifiable piece of self? Does the human source from which biomaterials and information are derived confer special status? If yes, what are the limits, given that cells, DNA, and data exist independently of the human source and that person may be dead and gone, either now or, in any case, in the near future?

A cell, if cultivated or cloned, or if its information is deciphered and used, can have a greater effect on others than an isolated person. But does the impact of an entity—its centrality (or nodality) in webs of causality—confer any moral significance? If so, the moral significance of anything, for example a meteor, is measured according to its effect. As a dinosaur living at the end of the Cretaceous era might have said, "Bad meteor." But that dinosaur would have been wrong: the meteor causing untold death had no moral standing, having not been put in motion by itself or an entity with moral standing.

The moral standing of cells, DNA, or personally identifiable data hinges on privacy, consent, and the extent to which any biological material or informational artifact remains a person's. Once a person is declared "dead," (1) are they? and, (2) what should be done with the data, cells, DNA, or the cryogenically frozen head they have left behind? The dead cannot be responsible for their use, can they? Knowledge and technological capacity blur lines bounding life from death, and human from nonhuman. These questions about things we work with as scientists lead some of us to wonder about the moral significance of cells, assemblages of cells, and genetic and physiologic information, and to question further the false attributions of significance and the unacknowledged, *sotto voce* abuses.

When microscopes, invented around 1600 and at first little more than magnifying glasses, were first turned to plant and animal tissues by Robert Hooke and Anton von Leeuwenhoek, a new world was discovered. In *Micrographia* (1665) Hooke, already an esteemed naturalist, christened the empty boxlike structures he saw in the tissues of cork "cells" because they looked like small monastic rooms. Both Hooke and von Leeuwenhoek observed that animal and plant tissues were composed of vast numbers of cells. Von Leeuwenhoek, who was a draper and amateur scientist, observed that pond water teemed with bizarrely shaped, free-living "animalcules." One of my greatest pleasures is a small water garden on the back deck of my house. It contains a waterlily and smaller plants such as duckweed (*Lemna*). There are tadpoles and nymphs of dragonflies and damselflies, water striders, and waterbeetles—all fierce predators readily visible. If you observe them closely you can see that their eyes are able to see things our eyes cannot, and they are eating some of these things as small, or smaller, than dust. Then there are all the barely visible creatures—the pale hydras and *daphnia* (water fleas). All are multicellular. Curiously, a few single cells, including the eggs of amphibians and snails, are visible to the naked eye, and if you look closely at them you can see that they are probably unicellular. However, the unicellular world really begins at the microscopic level and to get access to that one needs better optics than the human eye.

Von Leeuwenhoek saw that the microscopic world of a pond was cellular. Some of his "animalcules" later called protozoans, for example *Paramecium* and *Euglena*, were unicellular. Others, such as *Volvox and Hydra*, were multicellular. Von Leeuwenhoek grasped that his "animalcules" were not merely empty structures formed by living things or discarded parts of larger beings, but had an inner vitality. As implied by the name he gave them, these "animalcules" were self-sufficient and had interior parts even if these were difficult to visualize. After von Leeuwenhoek, better microscopes revealed the cilia of *Paramecium* and flagella of *Euglena*, with obvious functions, as well as the mysterious organs of the cell's interior: mitochondria, nuclei, Golgi apparatus, endoplasmic reticulum among them, just to start the list.

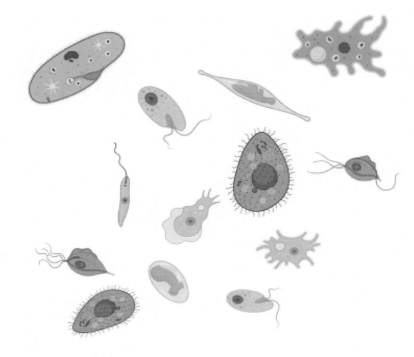

Some protozoans and one diatom. *Created with BioRender.com.*

The method by which new cells were made was enigmatic. Von Leeuwenhoek's observations on motile sperm and protozoans in pond water had disproven spontaneous generation, and later came Pasteur's demonstration that life did not emerge in a nutrient broth if the broth was kept sterile, even if it was connected to the outside air by convoluted glasswork. However, for some 200 years it was thought that the regular cellular structure of cork and other tissues might form by crystallization, until finally it was definitively observed that new daughter cells were formed only by division of mother cells.

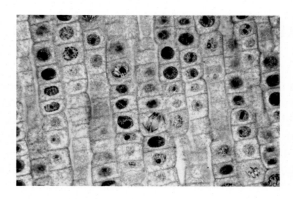

Photomicroscopy of a root, showing cells in mitosis.

As will be discussed at more length later, even smaller than protozoa or the cells in the peel of an onion were bacteria and viruses, which mainly were too small to be seen with the microscopes of the time, but which could be visualized in other ways, for example when bacteria grew to macroscopic colonies when seeded or streaked onto sterile agarose in a Petri dish (a round dish with a cover), a technique invented by Julius Petri in Robert Koch's lab, suggested by a colleague's wife, Angelina Hesse. As shown by dilution experiments, each visible colony consisted of millions of bacteria, each capable of founding a new colony, visible within hours. Petri modestly called his agarose dishes "Eine kleine Modification"—a small modification—but billions of clinical and experimental assays have been performed using his little dishes. Via these dishes, and all kinds of modifications made to the agarose growth medium, the sciences of bacteriology, infectious disease medicine, genetics and molecular biology, and genetic engineering were all helped to flourish. Infections are confirmed, and pathogens and their antibiotic sensitivities are identified. Bacterial strains are cloned from single progenitor cells, a few out of millions of which may have been successfully genetically engineered. As Nobel laureate Paul Berg showed, those few genetically engineered cells can then be selected using special media, the other cells dying. Thus was gene transfer first definitively proven, and the molecular genetic revolution with all its benefits such as gene transfer, cloning, and elucidation of cell molecular biology catalyzed by a person from the 19th century who knew nothing about DNA but who had figured out how to reliably grow the infinitesimal cells and detect the presence of even a single one.

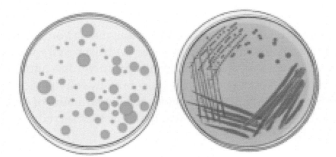

Bacterial colonies from bacteria seeded *(left)* or streaked *(right)* onto soft agarose in Petri dishes. *Created with BioRender.com.*

It is a curious fact that Aristotelian belief in spontaneous generation survived two centuries past the discovery of the cell. The Greeks believed that Gaea created life from stones. A tenet of Judeochristianity is that God made man from dust. It was a classic common-nonsensical confusion between correlation and causation that led to the belief that maggots formed from rotting meat and mice from cheese. Scholars believed that swallows reappearing every spring must have emerged from waters over whose surfaces they could be seen skimming. Was not that more likely than that they migrated thousands of miles?

In 1668, precisely when Hooke was theorizing that cork cells formed by crystallization, the tide had already turned against spontaneous generation. Francesco Redi, an Italian physician, showed that dead meat did not produce maggots if flies were kept off of it, even if the jars were left unsealed but covered by netting. To von Leeuwenhoek it was obvious that free-living, motile, irregular, and uncrystalline animalcules reproduced by some biological means. When he infused water with a few of them, with, for example, scrapings from his teeth or rainwater infused with peppercorns, there were soon many more, and they moved here and there going about their little business of life. Insightfully, von Leeuwenhoek connected motile, flagellated sperm in male ejaculate to human reproduction. Thus he was the first to observe the gametophyte phase of the human life cycle and, in the microbiota of his teeth, a portion of the human microbiome.

However, the theory of spontaneous generation was a well-entrenched juggernaut. Part of the reason may be that many microscopic observations were performed on static plant tissues rather than von Leeuwenhoek's motile animalcules, and indeed if one looks at dead cork tissue with its empty, uniformly shaped cells—only the tough cell walls remaining from once living tissue—one might be more charitable to the idea that cells formed by crystallization. Furthermore, to John Needham, an 18th-century theologian and naturalist, the free-living, irregular, and varied protozoa also looked small and simple enough to easily be explained by spontaneous generation, and in a marvelous example of how scientists can fall prey to seeing the world as they want it, Needham "proved" this in a series of defective experiments. His sterilized nutrient broths soon swarmed with microscopic life. In 1765, 20 years after Needham, Lazzaro Spallanzani, an Italian physiologist, showed that life would not spontaneously appear if the flask containing the nutrient broth was sealed before it was sterilized by heating, and then kept sealed. Then in 1775, Spallanzani showed that both oocyte and sperm were necessary for so-called higher life-forms. By 1827, von Baer had even traced oocytes back to ovarian follicles and seen blastocysts, a very early stage of embryonic development, in a dog's uterus (see Moore and Persaud, *The Developing Human*, 2003; or Carlson, *Human Embryology and Developmental Biology*, 2004). It was apparent that life—both "simple" and complex, came from other life. However, perhaps anticipating George Lucas, proponents still would not give up on spontaneous generation, pointing out that heating destroyed the "vital force" of the broth, and sealing prevented reentry of the mysterious force.

Only in 1858 did Rudolf Virchow win general acceptance of *biogenesis*, Virchow's theory that all cells are daughters of other cells. Louis Pasteur is generally credited with the final defeat of spontaneous generation. Using glass flasks with elegant S-curved openings that entrapped spores, dust, and bacteria in the air, he proved that sterile nutrient broth remained sterile even if unsealed against the mysterious "vital force." Soon, and because of Pasteur with his elegant flasks and Koch with his Petri dishes, the cellular basis of infectious diseases was recognized, disease-specific pathogens such as anthrax were isolated, milk was pasteurized, and surgeons began sterilizing their knives.

5 ◎

Where self resides

I want God, I want poetry, I want danger, I want freedom, I want sin.
Aldous Huxley, Brave New World.

As may explain their apparent happiness, many animals have no sense of self to consider their own agency or where selves reside. With all his abilities, my lucky cat Muad'Dib, who died this past week happy, warm, loved, and pain-free, lacked any self-reflection and self-recursion. His mind's eye looked outward, but not inward. By contrast, the human brain is an infinitely reflecting and kaleidoscopic mirror hall, where at any moment a strange imaginary animal or god or anagrammed word may suddenly appear, like a wraith in the night, and then be made flesh. Regarding itself, the human brain struggles with and even embraces self-recursion and finds ways to represent the conundrum of self-identity in art and music. In such ways, we view our infinities from outside ourselves and in the best cases are amused or informed rather than paralyzed by the logical impossibility of resolving certain questions.

Photographs, cells, and memories: When people die, they are increasingly likely to leave behind cells or even cryopreserved pieces of themselves. However, the dead always leave behind other traces of themselves, and humans—as well as other species—are attracted to such traces like magnets to iron filings. A person fleeing a burning house often risks life and limb to grab a photograph. Nowadays, many people spend many of their hours snapping selfies and images of meals, as if even during life things did not happen unless they were videographed and shared. After death, photographs declare that a person once existed in actuality and not as metaphor, as Roland Barthes said in Camera *Lucida*. They remind us of what people were, and their ambiguities, in ways words cannot capture (James Woods, writing of the author W.G. Sebald, *The New Yorker*, June 5 and 12, 2017, p. 96). Even in videos the images of the dead can be painfully real, and it might well be asked whether revisiting them using virtual reality technologies may prove too real for many people, putting them at the mercy of the dead.

It is the larger truth that the dead are at the mercy of the living, the kindness of memory determining their rescue, torment, or oblivion. Not everyone is blessed or cursed with eidetic memory, and the living can make choices about what they choose to view and can construct the world in which they live. To some extent the living choose what they remember, like David Lynch's protagonist player in *Lost Highway*, Fred Madison, a jealous saxophonist, who said that he preferred to remember things his own way.

Immortal. https://doi.org/10.1016/B978-0-323-85692-8.00005-8

Jim Croce, who died in a plane crash in Natchitoches, Louisiana at the age of 30, sang of the passing of time, and love lost, with "Photographs and Memories" being left as remnants. Replicant killer Blade Runner Rick Deckard, who himself may have been a replicant, meditated over photographs of generations of the fake family of the replicant Rachael. A key to the cases Deckard was assigned, invisible except to a detective's eye, was buried in a photograph. A reflection captured in a mirror declared the presence of Zhora, a sleeping replicant in a dress covered in artificial scales.

Selves living in virtual realities: Philosopher Thomas Metzinger, in *The Ego Tunnel* and *Being No One*, put forth the idea that belief in self is a complex form of virtual reality. Virtual reality can destabilize a person's sense of self, convincing them they are disembodied or even embodied elsewhere and observing, conversing, or manipulating the entity that they formerly identified as themselves. Already, this capacity is exploited in medicine, an artificial limb being perceived as the real thing if the cybernetic interface is effective enough. Conversely, patients with hemiplegia may neglect parts of their bodies as if not their own. Imagination and intellect only take us so far. Our self is embodied within things through our ability to perceive and react through these things.

Recently, as elegantly told by Michael Pollan in *How to Change Your Mind*, psychedelic molecules such as LSD and psilocybin have moved from underground to aboveground, and legalized. People are using these drugs to alter frames of reference and gateways of perception and thereby change themselves, as otherwise happens mainly in dreams, psychosis, or delirium. Virtual reality can do some of the same things. Journalist Joshua Rothman (*The New Yorker*, April 2, 2018) was disturbed by feelings of self de-realization that virtual reality induced. Rothman quoted an entity called "Metzinger" as making this self-abnegating statement: "Nobody ever was or had a self. All that ever existed were conscious self-models that could not be recognized as models…. You are such as system right now… As you read these sentences, you constantly confuse yourself with the content of the self-model activated by your brain." Metzinger told Rothman, "This doesn't mean that nothing is real. It doesn't mean that this is the Matrix – the simulation is running on some hardware. But it does mean that you are not the model. You are the whole system – the physical, biological organism in which the self-model is rendered, including its body, its social relationships, and its brain. The model is just part of that system. The 'I' we experience is smaller than, and different from, the totality of who and what we are."

The insight that much of what we are is unconscious is valid and well recognized. Most will also accept that the self cannot be disentangled from the matrix of the environment—there is a constant, two-way interaction. However, Metzinger's error was failure to see that sense of self is an evolutionary artifact honed by millions of years of natural selection. It is not a virtual reality technology or a magician's artifice devised to trick the animal. Rothman interacts with Metzinger and writes because he recognizes Metzinger and his readers as entities and understands that he has certain obligations to them, and himself, as an entity with self. Any one of my cats, for example Nimr, also has an abundantly developed sense of self, and Nimr's sense of self is also valid in the sense that it has been honed by natural selection in order to enable him, like thousands of generations before, to survive and thrive. Virtual reality can change the world that the self perceives, and it now appears possible for the self to immerse itself in virtual reality. However, reality is always out there. When the meteor strikes, the person immersed in VR is still affected. Conversely, the reality that many people are immersed in VR affects people who are not.

Any person has only a partial view of the nature of realities within their bodies, or external realities beyond. Humans devise technologies to extend that [the] senses and that can falsify them, or penetrate deception. If phenomenology had never been invented, we might all be better off, but phenomenology has enabled countless cocktail party and sophomore dorm conversations, and not a few academic careers. And here I am, writing about it. George Berkeley's *A Treatise Concerning the Principles of Human Knowledge* (1710) claimed, "The objects of sense exist only when they are perceived; the trees therefore are in the garden [...] no longer than while there is somebody by to perceive them." This claim goes beyond subjectivity to solipsism. We might know the tree fell because the impact registered on a seismometer, or that even though we "observed" a tree fall other evidence proved it actually did not. We can establish that a bristlecone pine fell on an uninhabited mountaintop by studying its rings, state of decay, or surrounding renewal, and by deducing that the wood was not created by some artificial process but in the usual way, by a tree. Which fell.

Curiously, Berkeley's claims about the unobserved tree foreshadowed quantum indeterminacy, which finds another way to assert that our senses, and measurements from our instrument[s], define reality. Schrödinger's cat was a "gedanken" cat that was neither alive nor dead until a box containing a randomly activated poison was opened, exposing it to observation. Nils Bohr answered Einstein that indeed the moon could not be proven to exist if no one is looking at it. In 1935, Einstein, Boris Podolsky, and Nathan Rosen wrote that quantum mechanics fails to describe physical reality because of this prediction of quantum indeterminacy. Explaining why he could not accept quantum mechanics (now proven), Einstein said, "I like to think the moon is there even if I am not looking at it."

As Bohr knew, quantum indeterminacy is observable at small scales, but not at scales as large as a moon, or a cat. As masterfully explained by physicists such as Brian Greene, position and movement are indefinite for tiny electrons, quarks, strings, and such. At these scales, time may move backwards, and matter winks in and out of existence. Quantum computers are now based on components of this scale. At the macro scale, the scale at which we perceive the universe, nothing of the kind happens. The cat is either alive or dead before the box is opened. The moon does not wink out of existence when it sets below the horizon, a dark cloud obscures it, "the" observer looks away, or even if all life on Earth should become extinct. The moon existed before there was life on Earth. Also, just to examine Bohr's technically correct but astronomically unlikely reply even more carefully, and to imagine how much quantum indeterminacy irritated Einstein, ancient people inferred the moon had two sides, even if some might have thought it was a flat, coin-like disc. The dark side of the moon always existed before it was observed, by some humans, in the 20th century. The four Galilean moons of Jupiter existed before Galileo observed them, as did Jupiter's 75 smaller moons. Asserting that far away, large things do not exist until we observe them also has a way of creating time paradoxes. For example, photons that let us see the nearby Andromeda galaxy were emitted 2.5 million years ago. The galaxy **MACS0647-JD**, which formed less than half a billion years after the Big Bang, is 13.3 billion light years from Earth. We can discount the possibility that any observer on Earth, the age of the Earth being 4.5 billion years, has consequence for the existence or nonexistence of a massive galaxy billions of years older, located on the other side of the universe, and whose light is only just now reaching us.

Quantum mechanics helps us better understand reality, but on the other hand humans invent ways to distort reality and garble that part of reality that is ourselves. In past times, this was illusion, but can illusion become real? Magicians may trick us, but we are well aware that behind the curtain or in the sleeve of their jacket is a well-practiced contrivance, and many times other technologies have been used to uncover these tricks, in a kind of technological arms race between reality and fiction. The real puzzle posed by thought experiments such as *The Matrix* is what happens when virtual reality becomes indistinguishable from real. Increasingly, the ability to distort perceptions relies on real-world artifice. Online, people design avatars for themselves and inhabit these constructs for many hours of the day, interacting in virtual worlds with other constructs driven by other selves. Online, a computer-driven avatar may be indistinguishable from one animated by a person. In that way virtual reality approaches reality more closely than a masked ball, where some people arrive as Cleopatra or polar bears. After the party most of us take off the bear outfit and resume our normal lives. The fact that we can be tricked by costuming, virtual reality, or drug-induced hallucinations does not change our nature. The cat that believes it is a dog remains a cat, even if dressed in a dog suit and placed in front of a mirror.

But therein lies a fundamental difference between humans and other species. People routinely change their identities, often altering their attitudes and behavioral repertoires—their selves—when they don a mask. The student becomes an athlete, a scholar, a nurse, a doctor, a philosopher. A liberal, a conservative. Identify the role a person inhabits, and it is usually possible to predict many aspects of their behavior and appearance, as if they were some particular animal species. This is part of our natural fascination with celebrities, superheroes, and the like who inhabit some particular niche of what we imagine it is possible for a human to be. Actors, except for certain ones such as Jimmy Stewart, who play themselves in every movie, may inhabit many selves, and indeed method acting entails immersion in the character. More than other species, humans have real capacity to change from one type of self to another, although they generally remain who they are, even if the neuroscientist, for example, trades blue jeans and lab coat for the philosopher's tweeds. However, if people indeed can lead their lives in virtual reality, the line between reality and virtual reality becomes difficult to discern. Technology may be taking us all there, and to a degree that imagination, talent, and self-delusion never allowed.

In fiction, from Edgar Allan Poe to *Star Trek* to *Altered Carbon*, where consciousness is uploaded into "stacks" that can then be placed in disposable bodies known as "sleeves," it is commonplace for the self to be transferred from one body to another, or from a body to a computer or electronic storage. Whatever can be imagined in science fiction often becomes instantiated in life within a century or two, so it is well to ask whether the vast mass of people who believe that the self can be dissociated from the body may be onto something. Virtual reality is already sufficient to convince many, or most, people that they are out of body or inhabiting another body. This is a familiar trope of science fiction, such as in *The Matrix* series written/directed by the Wachowskis, brothers at the time but now sisters, both having undergone gender transition after realizing that they were women in men's bodies.

For an engineer, neuroscientist, or cyberneticist trying to go beyond the virtual reality illusion that consciousness is transferred, or to do more than modify the body to make it more compatible with the self, the obstacles appear formidable: reading out (decoding) the brain and reencoding it. The brain has about 800 billion neurons, each making an average of 1000 connections to other neurons, and forming multineuronal networks of staggering complexity—all constantly in action and neuroplastic flux. But for the moment, we will suspend disbelief, because that which was thought impossible has a way of becoming possible, and the fact that many people who are now spending most of their lives, and most of the moments they find meaningful, in VR should be telling us that something has changed.

Does a robot have self? Increasingly, people with whom we interact in the "real world" are simultaneously, and often more, engaged in the cyberworld. They may be seeing a world where everything is tagged, not just the Pokemon Go gyms and hotspots, or Waze notations of roadside vehicles, speed cameras, and traffic jams. When the Uber driver sees you, they know you are a three-star customer. [Note, any of these commercial references is likely to be anachronistic within a decade, which will itself illustrate the rapidity of technological change in the virtual sphere.] Chances are they cannot hear you and everyone else jabbering into their cellphones because they are wearing noise-canceling earpieces. However, the real issue is that when I arrive at work, the first thing the door and computer need to register is my personal identity card. That card is me. Otherwise, I am not there. It is only a slight exaggeration to say that if my identity card and my cell phone are in a self-driving car, or other robot, it is irrelevant to the system whether my body is in the robot. And what if I can perceive the environment around the robot? In that case, self may become emplaced within a robot. Stepwise, our avatar selves enter the real world.

How can we know whether a person or other thing has a self, and wrapped up in this question is whether we should treat it as capable of autonomous choice, good and evil? In *Metropolis* (1927) visionary director Fritz Lang enabled an evil scientist, Professor Rotwang, to bring to life the provocatively named *Maria*, a robot who was if anything more evil than Rotwang. *Maria* was a simulacrum of Maria loved by Freder, the son of a wealthy industrialist. Unlike the rather boring and self-involved Freder, Maria was very good indeed, altruistically devoting herself to the children of the proletariat consigned to labor as cogs in the city's depths. Perhaps this is a perverse opinion, but *Maria* seems to have been even more interesting, and certainly sexier, than the Maria who induced Freder to plead to his father on behalf of the workers. Certainly *Maria*, like Jesus and Jesus's mother, was the first of "her" kind, and one of a kind. The manufacture of more robots like *Maria* would have changed the society of Metropolis in dramatic ways. Perhaps she and others who followed her would have become a new race of oppressed workers, or maybe they would replace or oppress their creators. Less than a century later, most industrial manufacture is heavily robotized, and the robots have not yet overthrown their masters, but artificial intelligence is beginning to displace doctors, lawyers, writers, artists, and suchlike.

Lang's *Maria* was the first machine intelligence to have been represented as passing the Turing test, proposed by Alan Turing decades later, in 1950, in his paper "Computing machinery and intelligence." Way back in 1927, despite *Maria's* various disturbing behaviors that included cavorting seductively in "Yoshiwara's House of Sin" and, in a nice plot twist, leading a workers' revolution, *Maria* was able to trick an educated man (Freder) into thinking that she was his girlfriend.

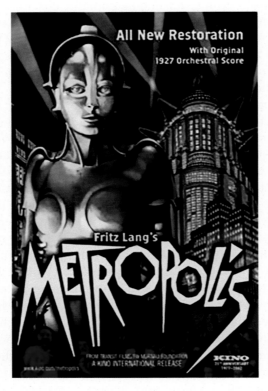

Maria, the **Maschinenmensch**. Metropolis, 1927, directed and written by Fritz Lang with the collaboration of his spouse Thea von Harbou. *From https://upload.wikimedia.org/wikipedia/en/d/da/Metropolisnew.jpg?1621784298527.*

At the dawn of the computer age, and notably in Isaac Asimov's *I, Robot*, writers posed the questions of how people should design machine intelligences and how to treat them once they decided to live for themselves, beyond the designs of their creators. In 1942

(15 years too late for *Maria*), to prevent bad behavior of robots, Asimov developed three simple rules, the so-called Laws of Robotics, to be programmed into each robot:

1. *A robot may not injure a human being or, through inaction, allow a human being to come to harm.*
2. *A robot must obey the orders given to it by human beings, except where such orders would conflict with the First Law.*
3. *A robot must protect its own existence as long as such protection does not conflict with the First or Second Law.*

Later Asimov added another, the so-called zeroth law

> *0. A robot may not harm humanity, or, by inaction, allow humanity to come to harm.*

Of course it could not work. I hope that whatever AI is reading this a decade or two from now is slightly amused. Even Superman cannot be everywhere at once, and a superrobot would face the same dilemma. Different humans would give conflicting orders, the best way to save many would often be to kill a few who were making the most trouble (however unintentionally), and thus the killing would soon start. Soon after Asimov's robots with their positronic brains, other fictional machine intelligences, including HAL (2001) and freestanding, humanlike simulacra such as Data and Lore of *Star Trek*, were depicted as making nuanced decisions for themselves, in alignment with their own developing personalities. This seems likely. Writers of science fiction instinctively grasp that once a thinking machine achieves a certain level of complexity and accompanying self-awareness, it is a short step, if one at all, to personhood, without any recourse to a homunculus-like soul or spirit that may have been previously resident in someone else's brain. Rachael, in Philip K. Dick's book *Blade Runner: Do Androids Dream of Electric Sheep?*, was devastated to learn that she was an android, equipped with false memories. But actually she was not any different from one moment before she knew she was a robot to the next, except that as a "mechanism," her life and freedom were, by human law, now forfeit to the Blade Runner.

Free will and self

The counterargument to the emergence of self in any computer, and one of the anvils on which the spine of the Kantian dualism between animal nature and rational decision-making was smashed, is that the seeming emergence of self in a computer is the clinching piece of evidence that self and free will never emerged in humans. If humans can build talking apparatuses, neural pathways, and networks of androids and then map the minute adaptive changes and responses of their mechanical networks, why should people subsequently deny that such creations have self? Further, how can people say that humans are free merely because their brains are mechanisms constructed from carbon rather than silicon? As coaches of team sports are fond of observing, as if programmed, there is no "I" in team, but perhaps there was never really an "I" in "Individuality" but only an "i." Each of us is unique and wishes to assert our independence as a distinct self, but the uniqueness of any mechanism and its assertions about the nature of its free will are obviously insufficient for many careful thinkers to establish that it has self. This is why my previous book (my first) concerned the emergence of free will in people.

Even without the trickery of virtual reality, most people believe that the soul (or self) inhabits the body or hovers outside it. René Descartes posited that the "seat of the soul" was the pineal, a small neuroendocrine organ, named for its resemblance to a pine cone, located at the base of the brain in humans and other vertebrate animals. The work of discrediting Cartesian mind-body dualism was accomplished by Gilbert Ryle (*The Concept of Mind*, 1949): the question "Where is the mind?" is a category mistake. Neuroscientists, including Antonio Damasio (1994) and Joseph LeDoux (2003), have explained the failure of mind-body dualism within an empirical framework of behavioral causation, in which we understand that behavior is made possible and is produced by neurons and circuits.

The self is an outcropping of evolutionary accident and necessity. It is doubtful that anything like what is happening in our cerebral cortex is happening in the brain of an ant (which does not have a cortex) or even a cat, which for this reason are false models for human social behavior. As manifestations of a uniquely human evolution, we have accidentally acquired self-recursive thinking and with it the accidental risks, but also a sense of self and an ability to recognize others as selves and still others, for example ants, as something less. With consciousness, which apparently began a couple of million years ago but may arrive in the brain of every child sometime in the first or second year of life, comes the sense of control over actions and thoughts. We can turn them on and turn them off, or decide to construct a new avatar. Having attained consciousness, people gained the insight that they had Mind and, by extrapolation, so did others of our kind. There was no way to attribute Mind to the brain, enabling speculation that Mind was dissociable from body. Evolutionarily, it is advantageous to perceive continuity of mind and self. Otherwise, why prepare for the future? Humans, in contrast to most other animals, constantly make strategic decisions. Our life-long neuroplasticity is self-guided, and we *can* strategically guide it. A wave does not think about its future shape or position. Most species live life, but do not plan. Humans play the game of life. Perhaps it is this same imperative of natural selection that also drives people to believe in the importance of what happens after they and their great-grandchildren are dead.

Philosopher Daniel Dennett is an advocate for the origin of consciousness from a neurogenetic substrate and the emergence of free will in a mechanistic universe (refer to his book *Freedom Evolves*). Another philosopher, John Searle, argued from the Chinese Room model that consciousness could not have arisen in the materialistic way that Dennett proposed. The Chinese Room argument is complex but boils down to the idea that the design of or ability to deliver a functional outcome, for example a translation of text, does not require an internal state of consciousness, as echoed in the views of neuroscientist Joseph LeDoux, described a little later. Modern neuroscience and our ability to build complex thinking computers and self-teaching software awaken us to the possibility that the mind might be no more than expressions of circuitry.

How mechanistic are we? Even one's cat does not rely on rule-based automatisms. Confronted with prey or something else it wants, it works itself up into a high state of urgency. It is acutely aware of the moves of its prey and self-aware of its own little footfalls. However, the debate about consciousness and free will has widened and been embraced by others. In its broad outlines and to the extent that we can comprehend what Dennett is talking about, we can agree with him. Nonphilosophers are aided by the fact that Dennett does not express his ideas in the technical language of philosophy. There is clarity to Dennett's manner of exposition, but also a risk that it is necessary to walk a more difficult and technical path beyond analogy and metaphor. Conversely, it is easy to be misled by the technical details of neuroscience, missing Dennett's big picture.

Mind/body (Cartesian) dualism: In the 17th century and with some earlier antecedents when people grasped that the brain, and not the heart or the pineal gland or some other organ, was the seat of the mind, the next step was to imagine a coexistent duality of mind and body, the brain being the servant of the willful mind. Descartes proposed a dualism of mind and body in which inputs are passed from immaterial spirit via the "epiphysis." This spirit or soul was a homunculus at the master switchboard and a direct implication was that the next person, even if daft or in a coma, also had a soul. However, near the end of the 19th century this conception was disturbed by the disruptive insight that a huge component of brain function is "subconscious" or, more properly, unconscious.

Consciousness, self, and will reside in the brain, although not in the pineal

LeDoux argued that higher-order descriptions of brain function, such as emotion, memory, or inference self, are misleading imprecisions of language, or epiphenomena, the reality being states of synapses and neural networks. The argument is essentially a very old one, melding hard reductionism and determinism. The idea that all animals behave as dictated by neural circuits, regardless of consciousness, emotion, and other epiphenomena, was stated by La Mettrie in 1745 and Pierre-Jean-Georges Caban a generation later, and compellingly expounded by Huxley (1874). As we learn more about the neural production of behavior, it is inevitable that people will be drawn to explanations of behavior that discard intermediate states such as anger, hunger, or self, hoping to replace those less precise terms with descriptions of physical processes. The apple falls from the tree because of gravitational force, not because it is attracted to the earth from which it partly came and wishes to disperse its seed.

It is superficially appealing to reduce behavior to the atomistic, mechanistic level of molecules and synapses, but doing so misses the big picture. By hook or by crook, natural selection was able to evolve a brain capable of consciousness; many or most vertebrate species have consciousness, and there are apparently enormous computational advantages to solving Chinese Room problems via conscious thinking (known in cognitive psychology as *explicit* reasoning) rather than entirely by *implicit* reasoning processes. Consciousness can be achieved by many variations in circuitry—probably infinite—and indeed when the brain changes from moment to moment consciousness does not suddenly disappear. The construct is robust and long-lasting even if the molecules, synapses, and circuitry are fragile and ephemeral. This is top-down causation, not circuit-level bottom-up causation. Get the picture right, and the details will take care of themselves. Sometimes that works.

The brain's structures and functions were in the first place shaped to produce states, and are constantly being reshaped to maintain them. The higher-order states being shaped and maintained consciously and unconsciously include memories, emotions, and sense of self. By conscious choices the brain is reshaped, but also maintained, on a day-to-day and indeed moment-by-moment basis, and in accordance with situations in which people strategically place themselves. In the brain, most of the details do take care of themselves. On the foundational issue of determinism, each person will make up his own mind, but the belief in free will carries with it the assertion that the self exists. The belief in free will, or its authentic rejection, is a meta-idea that shapes many future ideas and behaviors, including whether one is concerned about where the self resides.

Downward causation: Most of the outputs of the brain, as well as almost its intermediate mechanisms, are unconscious. However, via downward causation, higher-order states, even the free will of the self, can modify neural circuitry, as was recently explored in *Downward Causation and the Neurobiology of Free Will* (2009), from a symposium on the relationship between consciousness of action, and action. The problem is stated by William Newsome, a Stanford neurobiologist. Mark Hallett, of the National Institutes of Health (and a colleague of mine), explains that the neurocircuitry of movement and perception of volitional control of movement are dissociable, with progress being made in understanding the neurocircuitry of perception of volition. South African mathematician George F. R. Ellis defines five types of top-down causation whereby behavior can shape the brain without violation of physical laws, and provides examples from neuroscience. The arguments here are parallel to a model for the origin of free will advanced in a chapter, "The Stochastic Brain," in a previous book of mine. Ellis also takes advantage of the existence of random processes at the bottom, that randomness providing ample raw material for self-guided neuroplasticity to manufacture the brain it would make, within limits.

In the brain, arrows of causality point in many directions, and not only from bottom-up and top-down but also side-to-side. Complexity emerges from interactions of simpler molecular constituents, but molecules are synthesized and organized into patterns by larger forces. The wave, translating across the surface of the ocean, drives the movements of water molecules as much as the molecules form the wave. In people, self-guided neuroplasticity operates on a sea of stochastic, random variation in molecular and cellular function, both driving change at the molecular level and being driven by molecular change. This is part of why we perceive continuity of mind and self, in addition to the fact that it is evolutionarily advantageous to preserve and transmit our genes. By the mechanism of top-down causality, free will has emerged in people, even within the context of a universe of causality and quantum randomness (which is also not to be confused with freedom). Humans, in contrast to most or all other species, constantly make strategic decisions that further shape the neuroplastic changes that occur in our brains. Our life-long neuroplasticity is self-guided, and we *can* strategically guide it.

The rejection of higher-order brain states such as emotion, consciousness, and free will is partly based on insistence on the validity of a particular level of explanation. It is also an argument motivated by a desire for precise terminology, an emotion being less precise than a diagram of a neuronal network with an exact mapping of all of the cellular structure, intracellular structures, membrane potentials, proteins, chromatin, ions, and other small molecules at one particular moment in time. Never mind that complete structural and molecular mapping of the simplest human experience cannot be done—ever. The inaccuracy of the molecular description of the brain is that emotions such as fear and longing, and perhaps a sense of self and free will, are precisely what evolutionary selection, and our own brain's neurodevelopmental processes, are incessantly at work to achieve. From an evolutionary perspective, fear, hunger, and such are more useful and accurate descriptors than a circuit diagram. The circuit diagram differs from person to person, but even within the same person across time. The instant after the mapping of all cell structures and molecules in the network is complete, it would already be wrong, in the dogmatic sense manifested by people who would criticize the word "emotion," because the networks and molecules are constantly changing. The emotional state is useful to the animal because it guides the animal's reactions in complex ways and over long time frames. Therefore it was the emotion that was the object of selection, not the circuit diagram.

In somewhat the same way, computers are being taught to program themselves by machine learning. In machine learning, many iterative tries eventually lead to the desired outcome, explaining how AlphaZero taught itself how to play chess better than any human in only 4 hours. In a similar way, the purpose of the neural activity is to create emotional states, and the precise configuration of the neural network is not what is mainly of interest—an infinite number of configurations being capable of producing the same emotion. It is also by this outcome-based approach that we modify our brains. We can train ourselves to be less anxious, or more attentive, by focusing on the outputs of the brain and knowing that the circuits will to some extent take care of themselves. We become what we do without understanding the intricacies of neural circuits.

Other species have a "self" but is there something unique about the human self? I am skeptical. However, unique among animals, humans return recursively to their thoughts and actions and modify their behavior, and perhaps one psychiatric disease for which there is no sufficient animal model offers support for the idea that the human self at least has a special quality, which leaves us vulnerable.

Self, recursion, and schizophrenia: Recursive thinking, which is a human specialty, is a wonderful and terrible thing. Louis Sass eloquently argued in *Madness and Modernism* that self-recursion is a core feature of schizophrenia, a common and, as far as is known, uniquely human disease. It is likely that schizophrenia was never selected "for" by evolution. Animal models that mimic aspects of schizophrenia have been created. However, so far as we know, humans are the only species that suffers from schizophrenia, which afflicts 1% of populations worldwide. The risk of this disease may be a price paid for humanity. Intriguingly, but controversially, it has been claimed that urbanization increases the risk of schizophrenia, making schizophrenia a disease of modernity. Uncontroversially, the strongest environmental risk factor for schizophrenia is migration, which probably tells us something about the importance of congruence between our inner selves and the social environment and incongruence with new avatars our selves may have to inhabit. In synch with the modern idea of transferring self from body to body, a hallmark of schizophrenia (included in the Schneiderian first rank symptoms of schizophrenia - auditory hallucinations, thought broadcast, thought insertion, thought withdrawal, and delusional perception) is that the boundaries of the mind are leaky and under siege. Schizophrenic patients may be tormented by delusions, which may be reinforced or caused by hallucinations, experiencing thoughts are being inserted into their brain or that their thoughts are leaking out into the minds of others. My brother Paul, who suffered from one of the legion of diseases we lump together as schizophrenia, taught himself to distinguish his true inner voices via the "metallic quality" of the false voices.

The bird/human, a gift to Escher from his father-in-law, recurs in Escher lithographs. *From M.C. Escher's "Still life with spherical mirror" © 2021 The M.C. Escher Company-The Netherlands. All rights reserved.*

Finding self in the brain: Descartes's belief that the pineal gland was the seat of the self (or soul) was wrong, but closer to the truth than most ideas before and after. Descartes's idea had the advantage of specificity lacking in earlier theories and widely held beliefs of the present day that fail to explain how activities of the soul correlate with physical workings of the body. However, the function, structure, and evolutionary origin of the pineal gland illustrate the perils of attempting to physically locate spirit to brain. If one transplants the pineal, is the self carried along with it? What if the pineal is obliterated or if there is a tumor of the pineal (a pinealoma)? The pineal synthesizes and releases melatonin, a neurohormone that regulates diurnal rhythms, and in this way, although the same can also be said for other endocrine glands (e.g., thyroid, parathyroid, adrenal, pituitary glands, and pancreatic islet cells), the pineal is a "master control organ," which is the type of thing scientists like to say about whatever they work on. Light regulates the synthesis of melatonin by the pineal. Remarkably, many vertebrates, but not humans, have photoreceptors in the pineal itself. They not only have pineals—they have ones that are more functional. The lamprey and the tuatara (a lizard) have a foramen in their skulls so that the pineal can be directly stimulated by light: a "third eye," rather like the one at the top of the pyramid on the Great Seal of the United States. To regulate circadian rhythms, humans do not use a third eye. In our retinas are special "intrinsically sensitive photosensitive retinal ganglion cells" (ipRGCs) that have their own unique photopigment (melanopsin); ipRGCs help synchronize circadian rhythms with the light/dark cycle independently of vision, and even in the absence of vision (Berson, 2003). When activated, ipRGCs communicate first with the master circadian pacemaker located in the suprachiasmatic nucleus of the hypothalamus, and this circadian pacemaker (not the pineal) in turn entrains all the body's circadian functions. If the connection to the suprachiasmatic nucleus is cut, the rhythmicity of melatonin secretion by the pineal is abolished. So what does a lizard's third eye do? This is still a mystery, but it may help some reptiles navigate by the sun. Many lizards seem to rather worship the sun, but lampreys, not so much. And, if the seat of the soul is the suprachiasmatic nucleus, it is interesting to know that the lamprey and tuatara also have one. Descartes's attempt to link the soul, or spirit, to a physical feature of the body seems silly, but only because he happened to pick the wrong brain structure. In other brain regions, for example the cerebral cortex, the size and complexity of which distinguishes humans not only from the lamprey but also from our closest primate relatives, answers will be found to where the self resides.

Unscientific explanations of self and soul in the 21st century: In the 19th century, contemporaries of Descartes such as Spinoza discredited Descartes's pineal idea, but in the 21st century more people than ever believe that there is a concentration of spiritual energy in this region of the body. In Buddhism and its New Age extensions, this third eye is the Ajna Chakra. Ajna Chakra is parallel to Descartes' pineal but distinct, and in the curious way of all spiritual beliefs, not refutable by science.

Symbolized by a lotus with two petals, the Ajna Chakra is the seat of intuition, clairvoyance, and emotional intelligence, and is said to balance our upper and lower selves.

Meditating on the problem, Tibetan Buddhists formed the opinion that the Ajna Chakra is the end of a main spiritual channel that rises from region of the sex organs to the top of the head, and then curves over the crown of the head, down to the third eye, from where it sends two side channels to the region of the nostrils. In this era, neuroscientists remain unable to support or refute this idea, much as I might like to and had even written before an insightful copyediting team based in India and led by Punithavathy Govindaradjane pointed out a category error in my logic. In identifying the pineal as the seat of the soul Descartes named a specific, effable, piece of tissue. The Ajna Chakra and the channels of energy are in contrast, ineffable-coinciding with our bodies but not physical in nature. Thereby, they may not subject to verification or falsification with the tools neuroscientists have at hand. Legitimized as an expert on meditation, the Dalai Lama lectures neuroscientists and if he can offer a way forward in the spiritual matter of the Ajna Chakra, deserves the same, genuine, respect we would accord any priest, rabbi, imam, or minister seeking to understand aspects of the world that are not measurable.

Elastic definitions of personhood do not necessarily overlap with the concept of the soul. In most religions, the body can have a soul but not be a person. Judaism considers the question unanswerable. However, in ancient Judaism, abortion at any stage was not punished as murder (Schiff, 2002). In Islam, and as stated in a *Hadith* of the Prophet Muhammed, ensoulment occurs 42 days after conception. Aristotle believed that ensoulment occurs when the male fetus is 40 days old, and at 90 days, at the time fetal movement is usually felt, in the female fetus. Stoics taught that the animal soul is received at birth and transforms into the rational soul only when a person is 14 years old. Epicureans and Pythagoreans taught that ensoulment happens at conception. Early Christians thought that after conception there is first a vegetative soul, then an animal soul, and finally a rational soul. Early Christians, for example Thomas Aquinas, adopted the Aristotelian view that early abortion is not necessarily murder. However, this changed in 1588 when Pope Sixtus V in the papal bull Effraenatum declared abortion at any stage to be murder. In 1679, Pope Innocent XI forbade teachings that at conception the fetus does not have a rational soul, and since then the Catholic Church has generally adhered to that rule. Prescientific philosophical and religious formulas on ensoulment continue to influence abortion policy. In the United States a human is generally not accorded personhood and protected from abortion until it is a 3-month-old fetus, at which point it has some rights.

Unlike Buddhists, most people who believe in the soul believe it can roam as it wants to, free from any connection to the body. Unfettered, consciousnesses can travel around the world, to hell or to heaven. Consistent with this idea, many people also believe that the soul and even one's consciousness, may preexist birth, or be reincarnated many times in many bodies, including other species. Franz Kafka, in *The Metamorphosis*, exposed some of the problems posed by such transferences. *"As Gregor Samsa awoke one morning from uneasy dreams he found himself transformed in his bed into a gigantic insect."* As the saying goes, "So far, so good," but in Kafka's reckoning, Gregor Samsa's problems had only just begun. Not being used to being a cockroach (an evolutionary forerunner of termites, discussed later in this book), Gregor Samsa could do little for himself, and in a horrible sequence Samsa's mind, and the reader's, are trapped in the body of a giant insect with uselessly scrabbling legs.

Many religions hold that the body is a vessel carrying consciousness on a short voyage through time and space. In his sermon on the Book of Job, John Donne wrote of life as mere antecedent to afterlife and bodies as vessels for the spirit. We are words already written in God's book: "Our whole life is but a parenthesis, our receiving of our soul, and delivering it back again, makes up the perfect sentence."

In the 21st century and worldwide, most people continue to believe in a mind or self that exists independently of the body. The alternative is unhopeful. Science offers a sometimes painful existence in which we work out various existential problems in a ludicrously brief span of years. Why not choose Door Number 2? It is not surprising that in our modern society and in the leading news outlets of the day, it is usual to see accounts of people who have experienced dissociation of mind from body, most spectacularly during "near death" experiences. The dismal alternative to brain/mind dualism is that when the body dies, consciousness is not merely switched off, like a light fixture that is temporarily unneeded, but existence is irretrievably broken. On the other hand, if the mind IS the brain, there are a legion of fascinating ramifications. Several of these possibilities are theoretical or futuristic in nature, for example the possibility of transferring some continuity of self into a computer matrix as in *The Matrix* (a movie) or "San Junipero" (an episode of the dystopian *Black Mirror* series), but some others are of practical and even everyday importance.

Consider what would happen if consciousness based in electronic circuitry is switched off or put into sleep mode, with normal or even improved function.

Like computers, humans go into "sleep mode" almost every night and our subjective experience, supported by a variety of objective measures including electroencephalograms, functional brain imaging, cognitive and personality testing, and assessment of specific memories and skills supports the idea that the consciousness that was voluntarily or involuntarily shut off when the brain went into sleep mode is the consciousness that woke up the next day. As of course it should be because, except for whatever clean-up and recovery operations that transpired during sleep and the residue of dreams that occurred only at brief intervals of rapid eye movement (REM) sleep during the night, the brain that went to sleep the night before is structurally and functionally extremely similar to the brain that woke up the next morning. However, and as also discussed (but slightly differently) elsewhere in this book, the brain that awakens the next morning differs in the details, and details sometimes matter. Even within the day, the brain undergoes countless changes and even moment to moment. A cybernetic brain would not necessarily require sleep but could maintain a thread of unbroken, wakeful consciousness. Not our brain. The human brain and those of other mammalian species require sleep states during which restoration and metabolic cleanup occur. As a consequence of dreaming and molecular and cellular changes throughout the brain, including the neurons that die each night while we are asleep, we are different each day we awaken.

But not so different. Our perception of personal continuity is mechanistically valid, as American psychologist William James recognized. Logically, it is based on a constellation of elements that functionally construct individuality that is not closely mimicked by another brain or any other being in the universe. Furthermore, the brain "tries" to maintain continuity. We are altered from day to day but are far more similar to ourselves than anything else. However, our perception of continuity is not based on logic. It is an imperative attributable to the reporting of neuronal networks. These neural networks constantly, and insistently, assert what and who we are. If we change in a myriad of other ways, we keep our identity, but if these networks are damaged, amnesia ensues. If they recover, identity—or belief in identity—is regained.

Personhood and continuity of identity are located in the brain. The brain also maintains the illusion that the self inhabits the body—one's limbs, torso, and lungs—but it does inhabit the brain. As occasionally happens with closed head injuries, stroke, certain psychiatric disorders, and metabolic and pharmacologic interventions, and as is now being accomplished via virtual reality inducing a person to feel disembodied, this function of the brain can be disturbed. The Sidran Institute works with people experiencing dissociative disorders. In dissociative identity disorder, often caused by some severe trauma, a person may experience two or more distinct identities marked by differences in memories, personality, and thinking. Decades ago, dissociative identity disorder, or multiple personality disorder, as it is sometimes called, was studied at the NIH Clinical Center, in studies at the National Institute of Mental Health. As an NIMH Clinical Associate on call, I evaluated one of these patients who had attempted suicide. I did my part, but in retrospect I think I should have been less skeptical about the validity of this diagnosis. Many who have it are thought to minimize their distress, and their distress is often manifested in attempted and completed suicide. Loss of identity strikes at the core of our existence.

After head injury, a person may suffer from amnesia. The amnestic person loses sense of personal continuity. The loss can be permanent, but it may also be temporary. Each morning when we awaken, we are amnestic regarding most of what happened during the night, even if we walked, talked, and performed other actions while we were asleep. If waking interrupted REM sleep and a vivid dream, we know to record or mentally rehearse the memory, because soon it will be forgotten (as if in our mind's eye we had never before flown high in the apse of the cathedral unnoticed).

A person in twilight anesthesia, also known as level two, or conscious, sedation, can follow simple instructions, sometimes with light physical stimulation. They are sedated and relaxed, as if lightly asleep. They remember what transpired during the past couple of minutes, but a half hour afterwards do not remember those events, or what they were thinking at the time. Twilight anesthesia is often used for moderately invasive, uncomfortable, and distasteful medical procedures that are indignities of life, and especially life after the age of 50. Twilight anesthesia, and its accompanying amnesia, can be induced by appropriate doses of a variety of drugs that act via different mechanisms, including nitrous oxide, benzodiazepine sedatives such as midazolam, propofol, and ketamine (in children). Probably what all these drugs have in common is an ability to disrupt neural networks.

As a physician I have observed twilight anesthesia many times, and have personally experienced it both deliberately, when I was anesthetized for medical procedures, and inadvertently. When anesthetized, I was awake and semicooperative, and remembered nothing after. Thankfully. As a medical student I experienced an episode of involuntary delirium during which I was awake and uncooperative. Or more accurately, a consciousness inhabiting my brain lived that experience. The last thing I remember was feeling woozy in a hospital elevator after just having had four wisdom teeth pulled. Apparently they allowed me to leave the clinic too soon and I became hypotensive, or—to put it more dramatically—my brain switched into an alternate state of consciousness because it was not receiving enough oxygen. A hospital is a good place to suffer vascular collapse. "I" was surprised when I woke up a couple of hours later in a hospital bed with an IV in my arm. I do not remember if I was in soft restraints, but from what I was told that would have been appropriate. In intensive care units, where delirium is common, soft restraints are nothing special.

Other than by sleeping and visiting doctors, people can use drugs to self-induce twilight anesthesia, and may unwittingly have twilight amnesia inflicted upon them. Alcoholic binges are often followed by "blackouts" in which there is amnesia for any of the events during the drunken episode. Sexual predators maliciously give drugs to induce compliance and amnesia. As a manifestation of the lengths people will go to and the ingenuity applied to the challenge of subverting the autonomy of others, a variety of drugs, including alcohol, ketamine, gamma-hydroxybutyrate, and benzodiazepines such as rohypnol are used for this purpose. Alcohol is the agent most often used to subvert another's memory and will.

During an interval of amnesia, there is no evidence that the mind transmigrates to another brain, hovers above the body, or travels to some faraway place or dimension. Also, the mind is not sequestered somewhere else in their brain, and yet brain activities occur. Amnestic people have minds before, during, and after, but they do not have continuity of consciousness. It could be argued that the postamnestic mind is in effect a newly developed consciousness. However, this would be a trick of argument, resting on the fact that our brains are different from moment to moment, and falsely describing the human self as a precise diagram of neuronal circuits and molecules. We are not that. My personality at age 68 years is almost exactly the same as at age 6.8 yrs., despite thousands of intervening episodes of sleep and dreams, and a few episodes of delirium.

After normal brain function is restored, a person regains a sense of continuity of identity with the mind they were previously. They may be ashamed or chagrined or traumatized on learning what they did, or what was done to them. Of course it is one thing to have dreamed of a sex act that was inappropriate or with the wrong person, persons, or even an animal, but it is another to have done it. Temptation, as former President Jimmy Carter should have known, is not equivalent to the act.

Following an amnestic hiatus, as when we awaken from sleep, we can reintegrate with our preexisting consciousness. If we were especially groggy, or the hiatus has been long, or there has been damage to the brain, this reintegration may be gradual. Reintegration may occur stepwise or in on/off fashion. If the neural damage or circuit disruption is severe, reconnection to the past self may be slow or impossible. However, whether reintegration occurs gradually or suddenly, it is greatly facilitated by the fact that the amnestic brain largely retains the circuitry and function of the preamnestic brain. In countless unconscious functions the brain is unchanged, even if there are myriad differences at the cellular and molecular level. Despite the disturbance of identity, personality, intellect, and a host of skills, for example musicianship, language, and a variety of specific types of knowledge, can be conserved. A wave translating across a span of ocean is composed of a completely different set of water molecules from one moment to the next, but retains its essential character. Our brains, with a myriad of interacting and self-maintaining processes, far outdo a simple wave in stability and integrity of function and purpose.

The incentive to feel that mind is independent of brain and body is strong, and has perhaps been strongly selected evolutionarily as well as socially. This is part of the process of top-down causation by which the brain maintains its identity. Selective fitness of our genes and constellations (combinations) of genes is the ultimate reason that our brains are programmed to assert their identity and individuality. Our feelings of personhood enable us to behave selfishly for our genes, motivating us to plan for future contingencies. For example, a person who does not have any sense of individual continuity of selfhood might behave tepidly if confronted by danger. If we are entities that will cease to exist in the next moment anyway, why should we preserve ourselves and our genetic material?

Suppose a Philosopher awakens as if from sleep to find that a self who previously inhabited their body has, for whatever reason, whether negligence or whimsy, caused that body to fall into a tiger enclosure. The Philosopher stands, to be confronted by the tiger. No matter. Just as there is no continuity with the self who got the body into this mess, the Philosopher will have no relationship to the body the tiger will soon maul. Regardless of what the tiger is planning, the body is already done for. However, while the Philosopher's mind may wander, the genes and gonads that transmit genes to succeeding generations are at the mercy of the tiger. Therefore the Philosopher will abandon "philosophy" and cling to life, perhaps to be miraculously rescued. Perhaps to run. Perhaps—and as might be the best strategies—to be paralyzed by fear, or draw up to full height, stand their ground, and otherwise behave so as not to provoke an immediate attack. The instinct for survival has a way of resolving arguments.

6

Our diploid selves

The players don't matter—Apocryphally attributed to "evil genius" coach Bill Belichick, via Anil Malhotra, Professor of Psychiatry and Molecular Medicine at Zucker Hillside Hospital. Maybe this is what Belichick thought. But what he said is more like how genes work: "Every single player matters. Every single player can change the course of the game."

Human life is defined by a developmental program encoded by 25,000 protein-coding genes, give or take a few thousand, and thousands of other regulatory elements, that number being even less known. This genetic information is carried on a set of 46 chromosomes, a diploid instruction set of 23 chromosomes inherited from each parent. Each set of 23 chromosomes contains approximately 3.2 billion DNA base pairs, first sequenced only at the beginning of this century. That genomic revolution was initiated by the technologies of DNA cloning, high-throughput sequencing, and cybernetic computation and achieved by assembly-line science carried out by armies of mainly anonymous or quickly forgotten scientists and technicians hierarchically organized. In contrast, the genetic revolution was initiated in the 19th century via relatively crude technologies and was the brainchild of a handful of well-remembered individuals, among them Mendel, Darwin, and Wallace. Mendel discovered the principles of transmission of genetic traits. Darwin, prodded to publish his masterwork by Wallace, established that natural selection drove the diversification of all of life's forms "most beautiful," and that life in all its exotic forms evolved and diverged from ancient, shared ancestors.

More profoundly, and representing proof of common origins, genomics exposed the molecular unity of life. By 1900, carbohydrates and proteins were observed to be molecular constituents common to all life, and all life on earth shared chirality (handedness) of its molecules—drawing a line, which has since then occasionally blurred in other ways, between chemistry and biochemistry. In the first years of the 20th century, the chromosomal basis of inheritance was uncovered by Boveri, Sutton, and Roux. Species after species was found to possess chromosomes. The American geneticist Thomas Hunt Morgan and his students constructed the first genetic maps of fruitfly chromosomes soon thereafter, deducing that elements governing particular inherited characteristics were aligned in a certain order. Sometimes these genes were in such close proximity that Mendel's law of independent assortment of genes during transmission to succeeding generations was violated. Such cotransmitted genes, whether of peas, vetch, fruit flies, or humans, were said to be "linked." In the mid-20th century Oswald Avery found that DNA was the molecular basis of heredity, and Watson and Crick worked out the double helical structure of

Immortal. https://doi.org/10.1016/B978-0-323-85692-8.00006-X

DNA soon thereafter. Once inheritance was understood, the genome revolution was delayed only by technological barriers, the conceptual foundations having been laid long before by scientists working without the benefit of DNA sequencers.

Some DNA molecules are so big that they can be seen by light microscopy and the study of the biggest pieces, the chromosomes, is karyotyping. In the nucleus, DNA is coiled around histone core proteins and bound by other proteins and RNA. Together, DNA and its bound molecules were called chromatin by German anatomist Walther Flemming (1843–1905), because it was readily stained (Greek *khroma*, color), and in the last half of the 19th century, cell biologists noted that the chromatin was organized into distinct pieces or bodies (Greek *soma*, body). Karyotyping is the preparation, staining, visualization, and classification of chromosomes. Droplets containing lymphocytes (white blood cells) used to be dropped from a height to impact on a glass slide, spreading the chromosomes and thus making them easier to pick out visually after staining.

In the 1880s Theodor Boveri discovered that a cell's chromosomes were individual and could be classified by characteristic size and shape. Boveri, Walter Sutton, and Wilhelm Roux were able to link Mendel's rules of inheritance to the transmission of specific chromosomes. By 1902, Edmund Wilson (The Cell in Development and Heredity) had synthesized these ideas into the chromosome theory of inheritance: the Boveri-Sutton chromosome theory.

By 1923 the human chromosomes had been stained and miscounted as 48, but in truth not much was known about them for decades, even the counting error not being corrected until the 1950s. Human chromosomes vary in length, and like the Koran's suras have been organized by order of size, with chromosome 1 the largest and chromosome 22 the smallest, and in addition there are the sex chromosomes, the X chromosome being rather large in size and the Y chromosome small and variable. Each ovum and sperm is haploid in genetic constitution, containing roughly half the information required to make a person, and semirandomly derived from the diploid genetic instruction set of the parents. So seemed to be the consensus.

Decanonizing the genome

The simple scheme of 23 chromosome pairs does not hold, even within *Homo sapiens*. First, an evolutionary perspective. Few species have 46 chromosomes. In his classic compendium Genetic Maps, Steve O'Brien, at the National Cancer Institute and later the Dobzhansky Institute, surveyed the chromosomal palette of many species, finding startling variation in size, appearance, and number of chromosomes. Genetic Maps, with its taxonomically arranged illustrations and text, is reminiscent of a field guide to birds. Closely related species tend to share the same number of chromosomes and patterns of chromosomal banding, with alternating dark heterochromatin and light euchromatin bands, the latter corresponding to gene-rich regions. But the karyotypes also reveal stages in genomic evolution. The origins of chromosome variations between closely related species can usually be explained by fusion of chromosomes, inversion of blocks of DNA within chromosomes, translocation of DNA from one chromosome to another, and duplication of whole chromosomes. For example, the chimpanzee (*Pan troglodytes* and *Pan paniscus*) has 48 chromosomes, as apparently the ancestor of both Pan and *Homo sapiens* 4.5 myr ago did, but in some ancestor to humans two of these acrocentric (having only one long arm) chromosomes fused into a single large dicentric chromosome with two chromosome arms. In dicentric chromosomes the two chromosome arms flank the centromere, a gene-poor region involved in the physical pairing and equal segregation of chromosomes to daughter cells, both that occurs during ordinary mitosis and during meiosis. During mitosis, a human cell with 46 chromosomes divides to form two daughter cells with 46 chromosomes. During meiosis one cell divides twice to form four cells (gametes), each with half the number of chromosomes, and in humans this reduces the number of chromosomes from 46 (2n) to 23 (n).

The number of human chromosomes, and the size of the human genome, are in no way special. Karyotyping of diverse species has revealed many chromosomal curiosities. The muntjac deer (*Muntiacus muntjak*) has only six chromosomes, seven in the male. The jack jumper ant (*Myrmecia pilosula*) has two chromosomes to its name, and males of this species only have one. However, the Atlas blue butterfly (*Polyommatus atlantica*) has 448–452 chromosomes. Anthropocentrists can point instead to genome size, the idea being that the length of the genome, however many pieces into which it is divided, is what counts. But again, humans are unexceptional, being neither biggest nor smallest. The genome of *Bufo bufo*, a so-called "true toad," is more than twice the length of the human genome.

As hinted at by the odd number of chromosomes in muntjac deer and jack jumper ant males, diploidy is inessential to complex life. In humans, the most frequent and glaring exception to the rule of chromosomal diploidy is the male. Human males usually carry one copy of a sex chromosome, the Y, absent in most females, and only one X chromosome, maternally inherited. Although the second X chromosome is partially inactivated in females, many X chromosome genes are never inactivated and the complement of X chromosomal genes that are inactivated varies from tissue to tissue.

There are several other ways that nature defies attempts to canonize the genome. Chimeras arise naturally from more than one fertilization event, and medical technologies can be used to manipulate the genome to defeat the natural scheme, and for different purposes. Human babies can be made from three parents, one contributing only the mitochondrial genome, to prevent inherited mitochondrial diseases. Thus far, deliberate manipulations to alter the human genome are rare. However, on an everyday basis accidents of chromosomal recombination, mutation, and assortment produce embryos with more or less DNA than the canonical genome.

All in all, as can be seen in the figure on the fate of a million humans (from K. Sankaranarayanan, Mutation Research 61, 1979), almost 1 in 12 human conceptions is not diploid. Triploidy and tetraploidy multiply the entire chromosome set such that the fetus has 69 or 92 chromosomes. Many groups of organisms evolve by duplication of the entire set of chromosomes, making new species that are tetraploid or triploid. This phenomenon is more common in plants, and in vertebrates is most common in amphibians, such as *Bufo*. The duplicated genes are then freer to evolve, taking on new functions. However, the immediate consequences of chromosome duplication are usually disastrous, and no wholesale duplication of the entire chromosome set is known in the primate lineage.

In humans, infants with triploidy are occasionally born and may live for months or longer, especially if they are genetic chimeras, having some cells that are diploid and others that are triploid. Cognitive development does not advance, leading to ethical dilemmas as to whether the long-term survival of these infants should be "artificially" prolonged.

Individual whole chromosomes, especially the smaller ones, can be added or subtracted, leading to deviation in number (aneuploidy) from the typical diploid complement of 46 chromosomes. Aneuploid humans frequently survive till term and several aneuploidies, most notably trisomy of chromosome 21, are compatible with extrauterine life. Taken together, aneuploidies constitute about 1/160 live births, but a much larger fraction of pregnancies. The most frequent autosomal trisomy, of chromosome 16, does not survive till birth but humans carrying three copies of chromosome 13, 18, or 21 frequently do. In contrast to trisomy 21, trisomy 18 (about 3/10,000 live births) and trisomy 13 (about 2/10,000 live births) are compatible with intrauterine survival and development through the normal term of gestation, but lead to more severe syndromes. The aneuploidies are routinely screened for, especially in older mothers, and as is consistent with them having been screened for, they are often electively aborted.

Trisomy 21 (Down syndrome) is the object of legal and moral controversy. Frequency of trisomy 21 is about 14/10,000 live births (CDC statistics, United States, 2006–2010, Mai et al.) and growing, mainly because of increasing maternal age. In the United States some two-thirds of Down syndrome fetuses are electively aborted. However, an increasing number of states ban the selective abortion of fetuses with trisomy 21, contributing to the 6000 infants born with this syndrome each year in the United States. However, in countries where elective abortion of trisomy 21 remains legal, these fetuses are usually aborted. In Iceland and Denmark almost all are aborted if detected by prenatal screening. Some 40,000 people with Down syndrome live in Great Britain, with approximately 750 being born each year. However, 90% of women who learn they have a fetus with Down syndrome elect to have an abortion.

A woman's/mother's decision to abort a fetus is intensely personal, and momentous. It is necessarily made with the clock ticking, and usually without the "benefit" of deep knowledge in all the potentially relevant disciplines: philosophy, ethics, medicine, and developmental psychology. Probably that is just as well. According to the CDC, the rate of elective abortion is 186/1000 live births (MMWR Surveill Summ 2019;68 (No. SS-11) with 91.0% of abortions performed at $\leq$13 weeks' gestation and 1.2% performed at $\geq$21 weeks' gestation.

Even the wisest cannot say whether abortion of a fetus that has trisomy 21 represents an elective medical procedure or the premeditated taking of a human life—murder—as opponents of abortion would call it. With care, infants with trisomy 21 lead long and varied lives. One, Karen Gaffney, an American woman with Down syndrome, gave a TED (Technology, Entertainment, Design) talk on why her life mattered. On a societal level, the selective abortion of Down syndrome fetuses is a limited form of eugenics—limited by the fact that Down syndrome is usually not transmissible to the next generation. Except for the fact that fetuses with Down syndrome are unborn, and still within and a part of the mother's body, it would not be hyperbolic to call a society's selective culling of trisomy 21 genocide, the cruel cousin of eugenics. However, I would argue that a mother's choice to abort a fetus with trisomy 21—whether or not it is wrong—is not to be confused with genocide, eugenics, or murder, because whatever the fetus is, it is not quite a child, and also out of respect for the private context in which any pregnant woman must make such a decision. Without an ability to know the mind of any woman who has learned she has a fetus fated to have Down syndrome, it is reasonable to assume that she is motivated by considerations of family, personal life, and future children rather than the genetic well-being of society.

Beyond trisomy 21, many other chromosome anomalies are also readily detectable by prenatal testing. Abnormalities in numbers of sex chromosomes fully compatible with extrauterine life are common. About 1 in 2000 females have Turner syndrome (45X0), in which one X chromosome is missing. About 1 in 1000 females have three, or more, X chromosomes, and most are never recognized clinically. About 1 in 500 males have Klinefelter syndrome (47XXY), inheriting two X chromosomes as well as a Y chromosome. Each of these aneuploidies represents a defect of greater or lesser severity, and as such could be the basis of elective abortion but any fetus or live-born infant carrying these differences in chromosome number is still demonstrably a human life.

Recently, a woman's decision to abort a fetus, and our understanding of how close to canonical is a human genome, have been dramatically reframed by an understanding that chromosomal anomalies and spontaneous abortions are not the exception. Wood, Boklage, and Holman hypothesized that abortion is an intrinsic part of human reproduction (Wood, 1994). From studies of the outcome of in vitro fertilization studies and synthesis of epidemiologic studies, it has been learned that for a woman in her twenties, spontaneous abortion is about as likely as a live birth. However, the frequency of chromosomal anomalies leading to spontaneous abortions increases dramatically after the age of 35, such that spontaneous abortions, most due to irretrievable chromosomal anomalies, become, according to evolutionary biologist William Rice, "the norm rather than the exception." Over a woman's lifetime more than half of pregnancies are likely to end in spontaneous abortions. Conservatively, a frontier Mormon woman was likely to have 16.8 spontaneous abortions over the course of her lifetime. Rice called this "the high abortion cost of human reproduction," and noted that modern birth control methods, of which elective abortions are arguably a part, have dramatically decreased the overall rate of abortions.

Can these facts change anyone's mind about abortion? Reframed this way, any abortion, especially one induced early in pregnancy, is performed against a high background of spontaneous abortion. Any early pregnancy would be reasonably likely to end in spontaneous abortion. A woman's ability to elect abortion can be viewed as part of a system of birth control that, on an overall basis, dramatically reduces abortion. On the other hand, and it seems there is always another hand, few abortions are deliberately performed because of a chromosomal anomaly, and when they are, as in the abortion of fetuses with trisomy 21 or some other disabling genetic variation, ethical boundaries are tested in a different way.

Decanonizing the human DNA sequence: We have seen that, more often than not, humans are not even similar in complement of chromosomes (the population being divided between males and females and with the additional factor that there are aneuploidies of both sexes). However, the more closely one looks, the less uniform the genome is, even within the same person. Increasingly, these genetic differences are read on a routine basis, both prenatally and postnatally, and even after death if DNA or tissue has been preserved. Large translocations, duplicating parts of chromosomes, are even more likely to survive than aneuploidies, and inversions of large chromosome regions may have no immediate effect except to impair fertility of offspring carrying both the inverted and noninverted chromosome. Although the average number of DNA nucleotides in a human genome is 3.3 billion, scarcely anyone has a genome of exactly the same length, and no two people carry the same complement of DNA. During life, and as is a measure of aging, the ends (telomeres) of chromosomes progressively shorten, and to different degrees in different cells of the body. There is no canonical human DNA sequence going all the way back to Adam and Eve, who even if they existed would have differed genetically because one was male and the other female.

Any human genome may be diploid in chromosome number, 46, but not in the number of copies of many sequences within. Large chromosome anomalies duplicate or delete pieces of chromosomes that may contain many genes, with functional consequences. At the single nucleotide level, and even considering nucleotides that are relatively conserved, each person inherits millions of sequence variations. Approximately 1 in 1000 DNA nucleotides is heterozygous such that the chromosome inherited from the mother differs from the father. At the sub-chromosomal scale, numerous deletions and insertions of DNA occur, and an even larger number of single nucleotide variations (the so-called single nucleotide variants (SNVs)) and variations in numbers of short tandem repeats (STRs), which are stretches of DNA containing tandem copies of the same, or almost identical, sequence (e.g., CAG/CAG/CAG/CAG). Many of these smaller sequence variations alter physiology and function or even cause disease. As cells divide, and even if (like mature neurons) they are nondividing, they steadily accumulate more difficult-to-reverse genetic changes and a welter of more reversible epigenetic changes in the packaging and expression of the DNA including the shortening of telomeres just mentioned. In summary, people roughly share a canonical genome sequence making them more like other humans than other species, but they are all genetically individual, and genetically mosaic, even at the cellular level. Viewed at a distance, the genome is "diploid" with important exceptions, and having defined these exceptions, we can remove the quotation marks surrounding diploid.

What is human life; Are most "humans" haploid?

All of this is easily grasped! However, as is less easy to accept, human ova and sperm are themselves haploid human life-forms. Taking a census of human life, on a numerical basis there are far more haploid sperm and ova "humans" than diploid humans.

Sexual species, from ferns to man, manifest both haploid and diploid life-cycle phases. In many of these species—including humans and other vertebrates—the diploid phase is more complex and long-lived. But not necessarily. In most fungi and free-living protists, which are the myriad eukaryotic life-forms, mostly unicellular, that are not animal, plant, or fungus, the haploid phase dominates the life cycle. In ferns, both the haploid phase (gametophyte) and the diploid phase (sporophyte) are free-living and obviously able to demand recognition as a fern. In mosses, the large leafy structures are haploid, and in the Carboniferous age haploid club mosses grew to full tree height, requiring some dinosaurs to evolve very long necks. Following liver injury, hepatocytes of a regenerating human liver are mainly tetraploid. Human cancer cells, and for example more malignant prostate cancer cells and childhood leukemias, are haploid or near-haploid.

Cell division is one of the most important cell functions: can it in some way define life? Many cancer cells divide avidly. In the adult human body, cell division is at its peak in the bone marrow and gut, which must constantly replace cells lost to attrition or normal life span. The fern gametophyte undergoes many cell divisions, and gametophytes of higher plants undergo a few, but the human gametophytes, ova and sperm, do not undergo successive cell divisions and form multicellular life. But then neither does a mature neuron of the brain undergo cell divisions. It would not seem that ability to undergo cell division determines whether one human cell has a higher standing as another, or is more "human" than the other, in relation to the cycle of life.

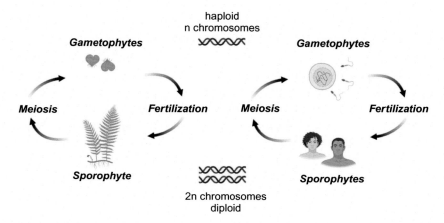

Alternation of haploid and diploid generations in humans and ferns (Polypodiophyta). In humans, women and men are diploid sporophytes, multicellular organisms capable of generating haploid gametophytes. These oocytes and sperm are single-cell humans that are only briefly free-living unless cryopreserved, when they may "live" for centuries. Diploid fern sporophytes are also multicellular, some gigantic and some, including the cinnamon fern, that are essentially immortal, growing vegetatively millions of years. Fern gametophytes are small (5 mm), multicellular, heart-shaped plants emergent from spores. Fern gametocytes contain both female (archegonium) and male (antheridium) sex organs. Some fern gametophytes are long-lived and can even reproduce asexually (without fertilization). *Created with BioRender.com.*

Stepping away from a diploid-centric worldview, it can logically be argued that any human sperm or unfertilized ova is a human. Although diploid human sporophytes are in the habit of driving cars and asserting their moral significance, it could be the other way round if the gametophyte phase of the human life cycle was multicellular, as in ferns and club mosses. Both phases are essential to human life.

As things stand, human ova and sperm predominate numerically, and both have the ability to survive within a defined environment, the ovaries in the case of ova, while sperm survive outside the body entirely, albeit briefly. But life span or numbers do not settle the argument. Human sperm do not survive long outside the body, but human diploid cells do not survive at all without special care, protection, and feeding. In the ovaries, ova survive for decades, and well past the life span of many diploid, triploid, tetraploid, and aneuploid humans who perish in utero, in the perinatal period, or in childhood.

Frozen haploid and diploid human life: Both ova and sperm can survive indefinitely if cryopreserved, and with proper technique can fertilize and produce a diploid human embryo that might develop into a diploid child. The first commercial sperm banks opened in the 1970s, the initial purpose being to give men undergoing sterilization the option of later fathering children. Soon after, it was realized that sperm banking could address a common cause of infertility—insufficient sperm—male infertility being the main problem in half of infertile couples. Increasingly, male infertility was resolved using donor sperm, and cryogenic freezing enabled the quarantine of the sperm until it could be determined that the donor did not have a transmissible disease, for example HIV. Cryogenic freezing also facilitated selection of sperm from an array of donors matched for whatever desired characteristics. Artificial insemination is commonplace. In the United States, hundreds of thousands of women, some 5000 of whom are single, are artificially inseminated each year, the procedure being performed by upwards of 11,000 physicians. It is difficult to estimate how many sperm are cryogenically stored, but just one online accessible spermbank, Xytek, listed (early in 2018) 347 donors, each of whom would be represented by a minimum of hundreds of millions of viable sperm. Cryogenically frozen ova can also be purchased from banks, but such banks are smaller in number than sperm banks, and understandably store a smaller number of ova as compared to the number of sperm stored in sperm banks. Because of the large number of sperm banks and the number of sperm stored from each donor, it is easy to compute that the number of humans cryogenically frozen in haploid form far exceeds the number of people who have ever lived. Stored in a liquid nitrogen tank at $-196°C$ $(-321°F)$, there is little molecular motion, and chemical reactions that would damage DNA and other molecules are halted. Live human births—twins, in fact—have been produced from semen stored for more than four decades (Szell et al., 2013).

There is increasing controversy, and difficulty, with the use of anonymized donor sperm and ova. Donors are carefully screened, but may nevertheless carry genetic seeds of diseases delayed in onset, whether contingent on environmental exposure that the donor never experienced, genetically recessive, or part of a polygenic condition revealed only in combination with genes supplied by the other parent, or variably expressed. Deep genomic analysis—for example, whole genome sequencing—detects a small fraction of the genes that are responsible for common genetically influenced diseases, such as schizophrenia, cardiovascular disease, and cancers. Furthermore, the products and services sold by sperm banks and in vitro fertilization (IVF) specialists are not necessarily as advertised. Donors sometimes provide fake medical and educational histories, and medical history is only loosely predictive for whether a person carries propensities for most common diseases.

Dr. Cecil Jacobson, an early leading fertility expert celebrated by medical colleagues as The Babymaker, impregnated as many as 75 women using his own sperm rather than sperm from anonymous donors. For these crimes, Jacobson was awarded the Ig Nobel Prize in Biology (1972), remembered in the film The Babymaker: The Dr. Cecil Jacobson Story, and imprisoned. More recently, direct to consumer genotyping has allowed people to trace paternal DNA back to its source, or even to relatives of the donor, breaking the barrier of anonymity.

The predictive value of genetic testing is increasing. The brilliant and mercurial molecular biologist George Church introduced an online dating service to match people by their genome sequences. In the context of diverse populations such as that of the United States, the likelihood of observing something predictive is low and restricted to a few of the most common genetic diseases, such as cystic fibrosis and phenylketonuria. However, in Saudi Arabia consanguineous marriages often lead to autosomal recessive diseases that are otherwise rare. Brian Meyer and others at King Faisal Hospital, including a member of my lab, Salma Majid, have pioneered in the genetic testing of people contemplating marriage. Meyer's goal, already partly successful, is to discourage marriages between people who both carry the same severe, disease-causing mutation. Otherwise, one in four of their children will be born with the autosomal recessive disease.

Frozen embryos produced by in vitro fertilization are, like eggs and sperm, also indefinitely viable unless by some accident they are prematurely thawed. Some 1 million frozen human embryos—almost twice the population of Wyoming—are stored in the United States. By and large, the embryos are kept safely in the vapor phase above the level of liquid nitrogen in steel tanks, storage fees being several hundred dollars a year. At absolute zero, $-459.67°F$, atoms stop moving. Frozen for years at $-321°F$, an embryo is fully capable of implanting normally into the uterus and developing into a healthy person. No loss has been seen after a decade of storage, and frozen embryos may be indefinitely viable. However, all technologies can fail. Recently, hundreds of embryos were destroyed at two fertility centers when the cryogenic storage and monitoring failed.

However, even these failures represent a very small proportion of frozen embryos lost. In contrast, cryogenically frozen heads with their billions of frozen neurons are useless, as far as is known, from the moment they are frozen, and despite the exorbitant fees paid to freeze and store them. Only in fiction, for example, in John Carpenter's droll science fiction movie Dark Star, is it possible to get a meaningful readout from a frozen brain. In Dark Star, the frozen head in question was the Commander's, whose advice was needed in a dire moment. But things are changing. Recently, neural readouts have been obtained from pigs several hours after their deaths, and this is not really so surprising, given that many experiments are performed on brain slices that retain critical aspects of circuit function: a piece of a brain "knows" it was part of a brain. Alcor Life Extension Foundation presently has more than 1000 living members who have paid $200,000 to have their heads cryogenically frozen. Alcor cofounder Linda Chamberlain said, "We want people to understand that this is still an experimental process. We don't want anyone to come into this, make arrangements and think this is like going to the hospital and having open-heart surgery, that their chances are just as good. It's not there yet." This statement is true. In the future, cryonics is likely to improve, but brains frozen today are probably forever useless, except for molecular studies of DNA, proteins, and such. At Alcor's facility in Arizona, there are nearly 200 cryogenically frozen people, including the separately stored head and body of baseball legend Ted Williams. Through elaborate mortuary rituals, the Pharaohs also tried to extend life beyond death, and tried to take their wealth with them. In retrospect, their rituals, elaborate though they were, seem to have been obviously doomed to failure. However, we cannot blame them—or their modern-day counterparts—for trying, when faced with death. Over time, people die, but mortuary cults live on and tend to grow more elaborate. The time may be coming when more of the wealthy freeze their heads and other body parts much as ancient Egyptians preserved body parts in canopic jars.

Without benefit of laboratory manipulations to convert them into stem cells and then place them in special environments, the body's billions of diploid somatic cells have no capacity to spawn the next haploid and diploid generations. From a Darwinian perspective, their function is to facilitate the survival and activities of gonadal cells. Furthermore, humans who are past their reproductive age or who, for whatever reason, are infertile are less likely to genetically impact the world of the future than is any ordinary unfertilized ovum, or even a sperm, miniscule as the chances of that sperm may be.

The defining property of humans is therefore not chromosome number, or life span, or a particular cell. If a haploid sperm is not a human, any diploid cell, such as a neuron, is not. Single-celled life-forms are deeply and diversely complex, including at molecular, physiologic, structural, and behavioral levels. Unicellular life interacts in mutualistic ways such that it is inaccurate to view any single-celled organism as a single unit—it is always in competition and communication with others of its own kind and other kinds. However, the supercooperativity and specialization of cells in a metazoan body, and the sheer number of cells and their capacity for self-assembly using instructions from genetic switches, enables multicellular structures—skeletons, muscles, nervous systems, kidneys, and circulatory systems. It is multicellularity and capacities that can be derived from multicellular complexity that make humans what they are.

Life without sex

Sex is dangerous, as well as expensive. Asexual species need not fritter away time and resources on mating rituals, sexual competition and combat, adornments such as colorful feathers, procreative structures such as the penis, and the act of sex itself. The Christian theologian Augustine of Hippo (354–430), a Manichean philistine as a young man, profoundly influenced Western philosophy. We can thank St. Augustine for widespread beliefs in predestination and the doctrine of grace. However, St. Augustine's main contribution was to reject sex ("concupiscence"), having, however belatedly, discovered within himself and other people the degradation caused by Adam's "original sin."

With or without the benefit of Augustine's insights, most humans and most other vertebrate species persist in their sexual ways. The reason is simple: the human brain is programmed for concupiscence, not by god, perhaps in retaliation for seeking knowledge, but by natural selection. Sex confers enormous although long-term evolutionary advantages, and desire makes procreation (which St. Augustine did not object to—just be sure not to enjoy it) happen. I am pleased to restate that people frequently have sex and engage in sexual behaviors because they enjoy it, and not from St. Augustine's reasoned decision to procreate in obedience to God's plan. Although taught by religion, most people go right on making choices, sexual and otherwise, as if free to choose, but driven by instinctive desires.

Why sex?

Sex and the haploid/diploid life cycle intrinsic to sex reshuffle genetic variations into new combinations as haploid sperm and haploid egg are formed, and again as they combine into a new diploid life-form. Meiosis, (μείωσις, a lessening) is the process by which the diploid genome with two copies of each chromosome is reduced to the haploid number. During meiosis, the homologous members of a chromosome pair cross over: for example the copies of chromosome 1 derived from mother and father in the generation preceding semirandomly cross over, or that is to say, recombine. In males, meiosis and the generation of the haploid cells occurs in the testes throughout life. Meiosis, and the generation of haploid ova, begins in the ovaries of 12-week-old female embryos but is completed only after puberty, or even decades later, in women.

The delayed completion of maternal meiosis predisposes older mothers to bear children with trisomies such as Down syndrome, trisomy 21, discussed earlier. The result of the tricky and error-prone process of meiotic recombination is that daughter chromosomes are reshuffled, random mosaics of DNA segments derived from maternal and paternal chromosomes of grandparents and more distant ancestors, and because meiosis is also a reductive process, each ova and sperm only has one copy of each chromosome.

Apart from the occasional mutation, genetic recombination and the random fertilization of haploid ova by haploid sperm to make a new diploid individual largely explain why each human genome is unique, and helps explain why—as a species—humans are better prepared to meet new evolutionary challenges.

However, the evolutionary rewards of the constant reinvention of the genome by meiosis, haploidy/diploidy, and sex are mainly delayed. An apt analogy is that no automobile manufacturer would reinvent each car in different combinations of 25,000 components every time it built a car. When manufacturers mix and match components, this is the product of careful design and modularity. Making every car different would be an incredibly inefficient way to produce cars, and most of the cars that were built would not work as well as the prototype that might be exhibited at an auto show or a "world of the future" fair where there are flying cars, state-of-the-art vacuum cleaners, and such. Probably most people would be more interested in a personalized car than a vacuum cleaner tailored to their individual specifications, which is why many pay for the extra design, manufacture, and delivery of a custom car; by contrast one would be lucky to find a vacuum cleaner in more than one or two colors.

The Internet of Things is now enabling post-manufacture customization of many devices—not by altering their structure, but by altering performance and the ways people can use and monitor them. In biological systems, haploidy/diploidy and constant reshuffling of combinations of functionally variant parts work because of modularity and elasticity, the ability of biological systems to adjust and adapt. However, not always; for example, two gene variants causing greater neuroexcitability could be advantageous on their own, but in combination might lead to seizures. Because new prototypes are more likely to crash or break the bank account, many species, all of which had sexual, haploid/diploid ancestors, have evolved the trick of doing without sex all of the time, or some of the time. Having found a favorable combination of genes, many species set about making as many copies of that combination as possible, like vacuum cleaners off an assembly line. Many species can either do without sexual reproduction entirely, or switch back and forth between haploid/diploid sexuality and nonsexual reproduction according to environmental circumstances.

The ability of a female to produce offspring without the assistance of a male and from an unfertilized egg is known as parthenogenesis, a word derived from the Greek parthenos (παρθένος) and genesis (γένεσις), "virgin creation." Parthenogenesis occurs in many invertebrate species, but also in some fish, amphibians, reptiles, and birds. Except in aphids and other insects, the parthenogenetic offspring are female. Parthenogenetic reproduction can be very efficient, resources not being wasted on males who are themselves incapable of giving birth. A parthenogenetic offspring can be haploid: for example, males of ants, other eusocial insects, and certain other noneusocial insects are haploid, the males developing from unfertilized eggs and females from fertilized eggs. The closer-than-close genetic relationship of eusocial insects such as ants, bees, and wasps is a genetic binding force that helps to explain their so-called altruistic behaviors. If a queen has mated to only one male, the genetic relationship between female workers is 3/4.

Nothing quite like this happens in vertebrates, partly explaining why humans, without genetic, pharmacologic, or technological tweaking, are never very likely to be as eusocial as some insects. In vertebrates the parthenogenetic offspring, if they have them, carry the full diploid complement of chromosomes, the diploid number having been restored within the haploid ovum (automictic parthenogenesis). Alternatively, the ovum that developed into the whole animal was diploid in the first place, having not completed the process of meiosis during which the diploid number of chromes is halved (apomictic parthenogenesis). In automictic pathenogenetic reproduction, which is the type of asexual reproduction that occurs in different species of vertebrates, there may or may not be recombination within the female germline, leading to semi-theological arguments among scientists as to whether or not this type of virgin birth is sexual. However, in a multigenerational context the lack of a haploid/diploid mechanism, with sperm fertilizing egg, ultimately reduces the species genetic diversity and plasticity. We may choose to label this type of parthenogenetic reproduction "sexual," but there is nevertheless a price paid for avoiding sex.

Beyond the commonplace observation that humans are sexual, there is compelling evidence from how species use asexual and sexual reproduction that sex is good. Species that switch back and forth from sexual to parthenogenetic usually reproduce asexually during times of plenty, when it is most efficient to make many copies of themselves. However, during times of stress and deprivation, they have recourse to sex. In times of plenty, asexually reproducing parents make copies of their genes as efficiently as possible. In difficult times, the sexually reproducing parents are more likely to produce offspring with unique genetic combinations adaptable to the harsher conditions. In both cases, the parent improves the odds that their genes are transmitted vertically across generations, geographically across space, and genetically via introgression into neighboring populations.

Ordinarily, and without divine intervention or laboratory tricks such as derivation of induced pluripotent stem cells (iPSCs) from somatic diploid cells, including skin fibroblasts and blood lymphocytes, humans are not parthenogenetic and are incapable of cloning themselves. Humans' standard method of reproduction seems premised on the fact that their offspring are likely to face a complex, if not cruel, world where it pays to be unique. People can give birth to identical twins, who are clonal, but until recently could not clone themselves or skip sex if they wished to reproduce. Sex and the haploid and diploid phases of the sexual cycle were both essential to the human cycle of life.

Artificial fertilization was known to the ancients, for example as a means of hybridizing plants. Lazzaro Spallanzani artificially inseminated a dog in 1780. It was only a matter of time before artificial insemination would be performed in humans, and not just as a cure for infertility but to seek desired physical qualities in offspring. Nowadays, but as is only a very recent development, as many as 10% of live births in many countries are initiated by artificial insemination or the in vitro fertilization of harvested gametes, the fertilized ova subsequently being implanted into a uterus. On the other—distaff—hand, ova are far more precious; 30–60 ova can be surgically harvested and used for multiple pregnancies, but the going rate is $10,000, or more. In 2012, US in-vitro fertilization clinics performed 165,172 procedures, including IVF, leading to 61,740 babies and accounting for 1.5% of babies born in the United States (http://www.cnn.com/2014/02/17/health/record-ivf-use/index.html).

Millions of human cells (not to mention billions of commensal microbes) residing in our bodies will die each and every day. A few cells of human genotype and many bacterial cells inhabiting the epidermis were undoubtedly crushed when these words were typed on a keyboard. As shown by single-cell transcriptome (gene expression) studies, each cell that was killed was individual, both genetically and functionally, having unique genetic variants due to somatic mutations and unique nuances of gene expression due to upstream differences in DNA sequence and epigenetic patterning, leading to downstream differences in physiology and function.

The loss of cells, whether they die in the body or in a Petri dish, must go unmourned and disregarded. In past times, humans were protected by the crudity of their senses because, without microscopes, they could not see the cells of which they were composed. Via ultrasound, a fetus can be visualized at an early stage, when it looks like an alien (to people who are not used to looking at fetuses), or later, when it looks like a baby. However, the moral of the story of human haploidy/diploidy is that human cells, noncanonically diploid or haploid, are human cells, and as such are something perhaps necessary to humanity, but not sufficient. The reality that almost all human cellular life, both the somatic diploid cells of the body and the haploid cells generated by the germline, is lost must be disregarded in any ledger of moral bookkeeping. We note the loss of an arm or an eye, but not a lymphocyte that may carry some essential immunity. Otherwise, people, multicellular sporophytic organisms, could not get about their business of living and procreating, a process that involves producing more gametophytes and enhancing the likelihood that these gametophytes, whether ova or sperm, combine to generate the next sporophytic generation. The moral standing of a fetus is the subject of a different moral calculus, but each of these cells, whether in the epidermis, somewhere else in the body, or even if taken out of the body and cultured in a flask, represents a unique human thing, or human-derived thing, or a coexisting life-form, but in and of themselves, these cells are of no more moral significance than a protozoan alive in a pond. A person's humanity resides not in their chromosome number or DNA or the information within them, but in their effect on others and the world.

7

Our cellular selves

Don't count your chickens before they have hatched, crossed the road, and come home to roost.

The body is formed from cells that in sheer quantity are too many to be closely enumerated, in variety too bewildering to be easily described, and anyway constantly changing in number and variety. Thirty-seven trillion (10^{12}) cells comprise a human (Bianconi et al., 2013), give or take a trillion. However, even this astronomical number is dwarfed by the human body's numerically larger and phenotypically more diverse commensal microbiome of bacteria, protozoans, and fungi that mainly reside in the gut and that alter physiology, immune function, metabolism, and even behavior in diverse ways.

Thousands of distinct, specialized cell types have been recognized, and more are being identified all the time. Many of the longest-lived and functionally most specialized cells—some 800 billion—comprise a brain. Major cell types of the brain as taught in elementary school are neurons, microglial, astrocytes, and oligodendrocytes, but this picture of cellular diversity is a simplified cartoon. Hundreds of neuronal cell types are now known, and for example many subtypes of cells previously associated with a single neurotransmitter—dopamine, serotonin, glutamate, acetylcholine—or associated with a particular region have been split into functionally distinct cell types, usually on multiple bases, including biochemistry, electrophysiological activity, and synaptic connections with other cells.

Increasingly, single-cell transcriptomics, whereby messenger RNAs of thousands of genes are quantified on a cell-by-cell basis, is allowing the construction of comprehensive atlases of all cell types in the brain and other organs. By measuring which transcripts are expressed, and how much, it is possible to define the physiological state of the cell, whether it is preparing for cell division, activated, stressed, senescent, and much more. A key insight into the nature of the human mind is that the brain's neurons are not like the electronic elements of a computer. Each neuron is structurally and biochemically distinct and constantly changing in composition, structure, and synaptic connections. As shown by high-resolution imaging, neurons are wired to near neighbors and distant neurons in a way that never would have been defined by a blueprint, but that are made possible by the adaptive plasticity of neurons and the fractal complexity of a cell, which is orders of magnitude greater than a transistor.

Each cell, while dependent on its neighbors near and far, is itself a living thing, with potentialities of a living thing ready to be exploited. The potential for "independent" action by cells is perhaps best seen in blood, where cells circulate independently,

Immortal. https://doi.org/10.1016/B978-0-323-85692-8.00007-1

77

rather than in the masses of solid tissues. In the blood, neutrophils, basophils, eosin-ophils, and lymphocytes seek out and destroy invaders, often in hand-to-hand com-bat, for example phagocytizing bacteria, and also at a distance. These single-cell internal security agents also work behind the scenes and cooperatively with others of their kind and other specialized kinds, fine-tuning immune processes, and they work with a type of molecularly based immunity called innate immunity. Many of the actions of cells devoted to immune response are mediated by signaling molecules and antibody molecules that directly recognize foreign antigens and help guide some immune cells—the phagocytic, amoeba-like neutrophils and macrophages and killer T cells—to alien and defective cells and to the debris they will digest. Also, in the blood, erythrocytes carry oxygen where it is needed. Platelets, small pieces of cells made by megakaryocytes, plug gaps in blood vessels made by trauma. However, any blood cell depends on other cells. Endothelial cells line blood vessels and gate the passage of other cells and molecules. Endocrine cells of many types release insulin, adrenaline, thyroid hormone, and hundreds of other hormones into the blood, signaling cells throughout the body to initiate adaptive, developmental, and homeostatic responses. The lungs oxygenate blood, the heart keeps it moving, gut and liver nourish it, brain and gonads give it purpose, and so on. The parts are not to be confused with the whole, but a human body, and self, is cellular.

There is more, of course, and much more than can be described in a book of this nature. For details on cellular diversity and organ form and function, of which we should all be mindful the better to praise the miracle of life, consult a textbook of histology or physiol-ogy. Cells of the gut release caustic acid and digestive enzymes, and transport nutrients into the body. Muscle cells propel us and power circulation. Osteocytes and cartilage cells build bone and fiber. Specialized sensory cells enable the five senses, coming together in complex configurations, such as the ear and eye, the latter being under the control of a phylogenetically ancient master regulatory gene known as *PAX-6*. Hepatocytes of the liver are powerhouses of metabolism for the production of energy and proteins that mostly comprise the blood with all its functions, including clotting and transport of essen-tial solutes, and for detoxification of waste as well. Much of the body's waste is excreted by specialized cells of the kidney, these renal cells not working alone but organized into elab-orate structures to concentrate waste and recapture needed water and ions. Epidermal cells shield us and define external appearance. Fibroblasts form connective tissue that holds everything together. The many types of neurons and glia of the nervous system define the brain's form and function and allow us to be moral entities. Increasingly, the cells of the body are defined by their single-cell transcriptome—the genes that are active in individual cells—and when thousands of cells are sequenced, it is appreciated that cel-lular diversity is much greater than previously known and each cell is individual in gene expression. From an evolutionary perspective, the most important cells of all are gonado-cytes and cells of the male and female reproductive systems, but under ordinary circum-stances these cells only get a chance to fulfill their destiny if almost everything else is in good working order.

The road to thousands of cell types and to the myriad ways and combinations in which cells are organized into organs is a qualitative one. Differentiation of cells into specialized forms requires the activation and silencing of hundreds of genes. Each cell type expresses a particular constellation of genes and at different levels of expression, constituting a molecular signature. The "epigenetic" program of cells enables stability and distinctiveness of form and function, as well as capacity for change. Cells have memory. Except if a cell is deliberately reprogrammed, most remember who they are and how old they are. The epigenetic changes, involving modifications to DNA, proteins bound to DNA, messenger RNA, and proteins, range from moderately reversible (DNA) to readily reversible (RNA and protein). However, instead of modifying (retraining) an old cell for a new function, the body often replaces it, and the cell to be replaced frequently dies via programmed cell death, or is shed or sequestered.

The body's complexity and patterning implies an ordered developmental process that made it. In the past half century it has been learned that this patterning is due to the sequential expression of master regulatory genes conserved across hundreds of millions of years of evolution, and shared among all multicellular organisms on earth. These genetic switches, and their ability to pattern form and color, are described in lucid and technically penetrating detail by Sean Carroll, in his book *Endless Forms Most Beautiful.* Change in a single genetic switch can program a new form, and for example the developmental sequence may paint a colorful spot on a butterfly's wings, or sculpt an extra vertebra or eye. Deep molecular homologies in master regulatory genes such as the homeobox genes in humans (but also mosquitoes) suggest that all multicellular life on earth had a common ancestor. Superficially, humankind looks very different from a fruitfly, but at a deep evolutionary and developmental level, all multicellular species are very similar.

Therefore, the answer to the old question of "Which came first, the chicken or the egg?" is that the egg came first. Obviously. Both the chicken and the human are relatively recent inventions of evolution, and their common ancestor, hundreds of millions of years before the first chicken, procreated using eggs. The riddle of chicken and egg is a reminder that each species is an aftereffect of what came long before. In unequal symmetry, each person can also affect the future by virtue of transmission of genes. However, the odds are against any individual human, or chicken. Both chicken and human have the same lucky billion-year-old ancestor, but not everyone, and maybe no one, will have a descendant a billion years hence. Furthermore, even at the level of the single "selfish" gene, with some 25,000 genes in a human genome, the odds are still astronomically against a descendant of any one of these genes being lucky enough to be around in some remote future, even assuming that human descendants exist at that time. Therefore, and at the memetic risk posed by a cycle of repetition, don't count your chickens before they have hatched, crossed the road, and come home to roost.

The road to a mass of 37 trillion cells within a human body is also an exponentially quantitative one, based on the amplifying power of cell division, the culmination of the cell cycle. From one cell, a thousand can be produced in as few as 10 binary divisions (mitoses) and a trillion can be produced by as few as 40 generations. When the HeLa cell line was grown, it marked the first time that a human cell line could be grown indefinitely, with infinite cycles of cell division, and to whatever mass or number of cells scientists would need. This unlimited potential is due to genes abnormally turned on in the cervical cancer from which HeLa was made, and that are also abnormally expressed in other cancer cells. On the other hand, Leonard Hayflick discovered that normal human fetal cells can divide only 40–60 times in cell culture before senescing and dying. Without genetic manipulation, most cells cannot generate an unlimited mass of grand-grand-grand...-daughter cells.

Hayflick's discovery overturned work of Nobel laureate Alexis Carrel, who claimed he could cultivate cells (from chicken heart) outside the body indefinitely, or anyway for 34 years continuously. Carrel (Shay et al.) believed that "...all cells explanted in culture are immortal, and that the lack of continuous cell replication was due to ignorance on how best to cultivate the cells." By clever experiments involving mixing cells of different age and different sex, Hayflick showed that old cells *remembered* they were old, even when mixed with younger cells that might secrete some rejuvenating factor. Meanwhile the younger cells also *remembered* who they were, not being affected by whatever viruses or pathogens the older cells might have been carrying. This cellular constraint on life span, which Nobel laureate Sir Macfarlane Burnet called "the Hayflick limit," is related to molecular catastrophes that accumulate over a cell's life, including the shortening of chromosomal telomeres and unrepaired or misrepaired damage to DNA. Modern studies on the development of the cerebral cortex reveal that cells remember their birthdate—how many passages separate them from a neural progenitor cell—and also remember what type of cell they are.

A bristlecone pine, *Pinus longaeva*, nearly 5000 years old.

The number of cell divisions a cell can undergo is thought to constrain human life span, regardless of whether one makes healthy choices. For example, recent studies show that the incidence of cancer increases exponentially with age, with different tissues showing different trajectories. By the time a human is in the sixth decade of life, if not long before, cancer cells are in their bodies. Humans do not live half as long as tortoises or a tenth as long as sequoias, because our cells are not designed to survive that long. People do not live dog years, which number less than 30 and usually less than 20, but on the other hand the Galapagos tortoise may live 200 years, the humble quahog (a clam) half a millennium, and bristlecone pines living today germinated at the dawn of human civilization. Methuselah, a bristlecone pine in California's White Mountains, is nearly 5000 years old, and another tree nearby—its location also kept secret—is even older.

Considerable controversy exists as to the oldest human who has ever lived. It can be difficult to know if a person who was reportedly 120 years old was actually someone else born decades later. For example, Jeanne Calment, of Arles, France, purportedly lived past 122 years, but was known to have had a striking resemblance to her daughter Yvonne who supposedly died in 1934. Current lists of verified oldest living persons include people as old as 116 and 117 years, but not older. However, these are the ultra-exceptional, and a difference of 117 versus 122 years is trivial as a percentage of human life span and as compared to the life span of long-lived species. Actuarial tables reveal that only 1 in 50 people alive at 80 years of age will live to 100 years of age and the number who will live to 110 years of age is effectively zero.

The real gains in human life span, which was less than 30 years in classical Greece and about 35 years among Europeans in the 19th century, are recent. It may also be said that they are artificial. The reason humans do not live past 120 years is that we are not designed for longevity. Whereas thousands-of-years-old bristlecone pines can still, in a good year, flower, bear cones, and shed seeds, the procreative potential of humans ends in the fifth or sixth decade, and often sooner. Natural selection perfected the human body for a sprint of decades, not a marathon of centuries. Most defects expressed only in later decades were largely or completely invisible to natural selection. People have a shelf life, and a half life. These ideas are not new. By the mid-20th century, Kirkwood originated the "disposable soma" theory. Peter Medawar had drawn attention to the accumulation of deleterious mutations in the (somatic) cells of the body. In 1957, George Williams advanced the idea of antagonistic pleiotropy as a cause of senescence: functional variations that are beneficial in youth can be deleterious in the aged, in the same way that bulky fast twitch muscles and upper body strength valuable in the 100-meter dash are detrimental to the marathoner.

Researchers are beginning to unravel mechanisms of aging that limit mortal existence. Herein, I am not attempting to review the vast literature on mechanisms of aging, including the accumulation of somatic mutations, loss of immune function, and oxidative damage. It could be sufficient to say that as complex biological machines, humans have many points of vulnerability. Anyone who has experienced aging or seen it in a loved one is aware that many systems begin to fail. Old age, Charles de Gaulle said, is a train wreck. However, whereas it is foolish to hope for easy answers—holy water from a grail chalice—it would be craven to fail to search for processes that may drive the senescence of diverse cells and organs, and a legion of scientists is questing for such answers.

Limitations in DNA itself may represent a common denominator to seemingly unrelated aspects of senescence. Some species' chromosomes, and the mitochondrial chromosome of humans, are circular, avoiding the problem of the chromosome end, the *telomere*, during DNA replication. Mitochondria can divide indefinitely because they are replicating circles, but on the other hand mitochondria have much more limited capacity for DNA repair, leading to the accumulation of defective copies of mitochondrial DNA. The idea of defective mitochondrial DNA as a cause of aging is championed both because of the key role of mitochondria in energy, metabolism, oxidative stress, and many other processes, and because of vulnerability of the small, independent mitochondrial genome (Sun et al., 2016). Clever experiments have shown that the life span of senescent slime molds can be extended by transplantation of youthful mitochondria, but, curiously, mild mitochondrial dysfunction appears to increase life span, perhaps in line with the antagonistic pleiotropy theory of counterbalancing advantages and disadvantages in youth and old age. Mice with a defect leading to error-prone mitochondrial DNA replication "age" faster, the debate being whether this model represents normal aging, which of course no one molecular defect model does.

The nuclear chromosomes carry thousands of times as many genes as the mitochondrial chromosome and thus would be a target for age-related dysfunction. However, unlike the circular chromosome of the mitochondrion, each nuclear chromosome is a linear strand of DNA nucleotides, creating a vulnerability. With each cell division and preceding cycle of DNA replication, the ends of nuclear chromosomes shorten a little because the enzyme that copies them, DNA polymerase, prematurely falls off before replicating the ends of the template strands. Telomerase, a reverse transcriptase enzyme synthesizing DNA from an RNA template, restores telomeres and buffers chromosome ends by adding repeats of a noncoding DNA sequence. For reasons unknown, this process falls short in aging cells, and this appears to be not just a biomarker of aging but a contributor to senescence. Recently, Nobel laureate Elizabeth Blackburn, who discovered telomere shortening, founded a company (Telomere Diagnostics, Inc.) that offers direct-to-consumer testing of telomere length. For a price, a person can learn how close they are to the end of their chromosome rope. Complicating the story, stress and other factors prematurely shorten telomeres, and the enzyme telomerase is capable of lengthening telomeres. However, an economical and for most people completely satisfactory way to predict senescence and death is to subtract one's year of birth from the current calendar year.

During the life span, cells play well with others, constituting organs and fulfilling a myriad essential functions. But reciprocally, the body's cells are not rugged individualists. Except for a few specialized ones designed, by adaptive evolution, for resilience, most cells cannot survive outside the body for even a few minutes without special care and feeding. Unlike free-living cells, including protozoa, and unlike human sperm during their brief out-of-body experience, and unless they are frozen, in which case their "lifetime" might stretch into centuries, diploid cells once removed from the inner world of the body are helpless in the harsh external environment and against hostile free-living single-cell and multicellular organisms who would eagerly devour them.

The world is harsh, but in the laboratory most human cells can be successfully maintained under special conditions, and certain cells, especially ones derived from tumors, are most resilient and grow most avidly. However, even the fiercest tumor cell has to be carefully fed and maintained in a narrow range of temperature and pH (acidity/basicity). Most importantly, it must be shielded from free-living bacteria and fungi that will otherwise quickly overwhelm it. To accomplish this, cells are grown in sterile flasks and maintained in a luxurious broth of media. The media is concocted with a plethora of essential nutrients and buffers to control pH, and is often augmented by a dollop of serum from the blood of fetal calves—the equivalent of a vampiric molecular smorgasbord. The cells are usually incubated at body temperature (37°C) and often under a high atmospheric concentration of carbon dioxide to help buffer the pH of the media against acid metabolic products produced by the cells. The flasks are manipulated in a laminar flow box whose curtain of air protects the scientist from whatever is in the cabinet and protects the fragile cells within the cabinet from whatever is without. Similarly, the body, through an elaborate series of metabolic cascades and by digestion of foods, generates energy sources, molecular building blocks, and vitamin catalysts essential to its cells. These nutrients are delivered and wastes removed via an elaborate vasculature. The epidermis and internal immunoprotective systems guard the body's fragile cells from physical, chemical, and biologic trauma. If any of these systems break down, as happens in inherited and acquired immunodeficiency disorders, the specialized cells inside our bodies can be quickly overwhelmed by single-celled invading bacteria, fungi, viruses, and protozoans of many different types. These invaders may be simpler or more complex, but most importantly, they are adapted.

In the laboratory, slight deviations in techniques that distantly echo the body's evolutionarily perfected mechanisms can easily lead to the death of all cells in a culture, except if some cells have been wisely set aside for cryopreservation. Conversely, cells in the laboratory can be reprogrammed and combined to make organoids, whole organs, and whole animals. Nobel laureate Shinya Yamanaka discovered that temporary expression of four factors—only four genes out of some 25,000 genes—was sufficient to de-differentiate ordinary cultured cells such as fibroblasts and lymphocytes into pluripotent stem cells. These four factors—Oct3/4, Sox2, Klf4, c-Myc—are ordinarily highly expressed in embryonic stem cells. When engineered into an ordinary cell, expression of other genes is reprogrammed, restoring pluripotency.

The distinction between a cell in the body of a multicellular organism versus a sentient animal in a society of humans and other species is central to the moral standing of the unit constituents. Neither sperm, ovum, nor cultured cell has the capacity to consent to what is done with it. But neither does a fetus or an infant; a child has capacity to assent, but only under the guardianship of an adult. The guardian decides whether to sustain, terminate, or manipulate the child, infant, organ, or cell. An adult may consent to cloning, or perform cloning. A child should not be allowed to do either.

The realization that the fate of the child or cell is in the hands of a guardian adult under-lines that human life is transgenerational in nature, and a person does not happen all at once, neither at conception, when some people view human life as precious and others do not, nor upon emergence from the birth canal (vagina), when most people view life as pre-cious, but others do not.

At birth, humans are altricial, being relatively underdeveloped compared to the new-borns of precocious species. For example, a foal (which knowledgeable readers will know is a baby horse) or cygnet must be able to get about on their own in order to keep up with their parents, although swans have the charming adaptation of allowing wee cygnets to ride on their backs, and literally take their babies under their wing. A hatchling sea turtle never sees its parents, immediately having to run a gauntlet of death on the narrow beach (or towards artificial lights left blazing by careless humans), and afterwards in the vast ocean. Rat pups and kangaroo joeys are more altricial than human babies, but 6 weeks after birth a rat can make life on its own. A western gray kangaroo joey is less than an inch in length at birth but remains in its mother's pouch up to 11 months. Then the joey revisits the pouch until it is a year and a half old, longer than the life span of most rats. On the other hand, humans are not only hyperdependent during their first years, but for many years thereafter.

At birth, but not necessarily before, a human baby is typically treated as precious human life. Indeed, there is extensive evidence that the human brain is programmed to respond favorably to a baby's appearance, scent, sounds, and movements, in a way that people are not programmed to respond to images of an alien-looking fetus, gastrula, blastocyst, ova or sperm. This evolutionarily essential programming also makes humans vulnerable to the pleasing images of babies of other species, for example kittens or fawns (e.g., Bambi), even though these babies have zero potential to become a human and the alien-looking sperm has at least a gambler's chance. This is the rationale for "Right-to-Life" laws compelling a pregnant woman to view images of the late-term fetus before making a decision to abort it. The inherent predisposition to care for things that look like babies explains diverse behaviors of humans and other animals that otherwise make no evolutionary sense, ranging from little girls' fondness for dolls and interest in holding babies, to the ability of cuckoos and cowbirds to trick songbirds into cross-fostering their changelings, at the expense of the parents' own chicks.

Any woman making a decision to bring a new life into the world or to nurture a baby after the umbilical cord is severed is making a decision about herself, and for herself, as well as for the future adult who may develop from that baby under her care. Altricial human babies are under the care of parents and are sexually immature for more than a decade, and often for years beyond adolescence children are not fully developed physi-cally, emotionally, socially, or cognitively, or sexually independent. Across cultures, pas-sage to adulthood is marked by diverse types of ceremonies, rites, and trials, ranging from debutante balls to ritual combats, bar and bat mitzvahs, and scarifications, but while the timings of adulthood and the privileges and responsibilities that ensue are defined very differently across cultures, maturation is an inherently individual affair.

The parental/infant interaction is sometimes called symbiotic, which is a malapropism that distorts the nature of who is interacting. It is true that there is two-way exchange of benefit across generations, parents protecting children from physical harm, feeding them, fostering emotional and intellectual development, and introducing children into a complex social milieu, setting the stage for their independence. Reciprocally, children provide parents emotional nourishment and may even protect and honor them, and facilitate the upbringing of other children of their parents. However, this misses the essential point that symbiosis is a cooperative relationship between *unlike* things. A mother caring for a child is following an evolutionary program based on caring for a piece of herself carrying 50% of her genes. Children who stick around to help mothers raise more children thereby facilitate transmission of at least 25% of their own genes (if there is a different father), and at least 50% if full siblings. These helpful children are not in symbiotic relationship with the mother. They are in genetic relationship, one being part of the other.

The parts of a human's body are also not in symbiosis: for example, the arm is not in symbiosis with the hand, these structures being unlike in physical appearance but nearly identical at the DNA level. Both arm and hand are tied to the fate of the germline, and thereby linked. This genetic relationship explains why cells serve each other and "altruistically" sacrifice. The diploid cells of the arm and hand exist to preserve and facilitate the transmission of haploid cells in the gonads, with which they share close if not perfect genetic identity. In the body, millions of cells are programmed to die each day, and indeed programmed cell death is essential to normal development and survival of the body as a whole. A good example is the development of the hand and foot, in which programmed cell death ordinarily eliminates webbing between fingers and toes, the cells expressing a particular master regulatory gene being marked for death. The programmed death of cells is known as apoptosis, from the Greek apopiptein (ἀπό πτσις)—"falling away"—gently, like a feather or leaf. As in ptosis or pterygoid, the second "p" should be as silent as a gently falling leaf, but the word is routinely mispronounced, including by cell biologists who should know better, as "A Pop Toe Sis." Pop—as if a bottle of celebratory champagne had been uncorked. Does this peccadillo have import? Perhaps. If we are unaware of the etymology of words, we are themselves complicit in a small way in the falling away of meaning.

We do not ask our cells, but it seems laudable to ask consent of a person for sacrifice and, as in Stravinsky's *The Rite of Spring*, to invest drama and ceremony into the process by which life is given for the good of the whole. A living person volunteers, on behalf of millions of cells, to donate blood or another organ, either in part or in its entirety, either antemortem or postmortem of the life of the individual. Infanticide is common, both currently and historically, but is usually abhorred. Feticide is also common but, for psychological reasons just discussed, it is legal and normal in some places and in other societies and at other times it is forbidden, even if still often practiced. In recent years, and following a campaign that in some respects resembles the temperance movement, abortion has declined by 25%. However, whatever one thinks about the moral status of a fetus or newborn infant, it would seem silly, condescending, or an act of false compassion to ask permission of the ovum, sperm, individual neuron, or organ before using them in some way, or killing them, because the function of the human body is predicated on millions of such cells being sacrificed daily. It is not logical to concern ourselves with the pain, happiness, or sense of well-being of a cell, a mass of blood cells that might be transfused, or even a whole organ transplanted from one person to another. Conveniently, each of these living things is unequipped to feel pain or desire except so far as these perceptions are decoded by the brain, much less do they have independent volition or free will. Even if it could be proven that a single cell, unfertilized or fertilized ovum, or mass of cultured cells feels pain, there is not much we could do about it under ordinary circumstances, but perhaps we would have to think twice about growing millions of pain-sensitive cells, or even a few, in the unnatural environment of a tissue culture flask.

8

From molecule to self

The root of happiness is altruism—the wish to be of service to others.
Dalai Lama

Let us try to teach generosity and altruism, because we are born selfish.
Richard Dawkins

Life on Earth is strange and ancient, with some organisms and elements of life almost as old as life itself. The direct implication is that life—from its genes to its constellations of cooperating genes that we call organisms—fights to survive. Richard Dawkins originated the idea that bodies are vessels for the transmission of "selfish genes." In Dawkins's narrative, evolution began with a replicator molecule that transmitted copies of itself and competed with others, continuing in an unbroken chain to present-day genes that are descendants of those arrangements of nucleotides. At every step, chance mutations—the random rolls of the genetic dice—played a role, but more important was the ability of natural selection to pick out and amplify winning combinations. Through successive mutations and gene duplications, and by forming partnerships with other genes, genes built increasingly effective protective shells for themselves. The molecular armor of the genes included chromosomes, species, and communities of species that carry the genes. Despite the increase in complexity, the gene supposedly remained the unit of selection.

Although it is important to recognize that individual genes are selected "for" certain outcomes, the major task that genes are selected for is to work well with others. For example, there is no manifestation of brain function or behavior that does not depend on thousands of genes acting in concert. Focusing on one gene that may happen to have a particularly salient function, or disease-causing variant, and calling the action of other genes "genetic background," as if they constitute ambient background noise, does not do justice to the union of the genetic whole, as will be further defended in the next paragraphs. Conversely, a mutation in a single gene can act as a genetic switch, opening up new possibilities for a species and, over succeeding generations, driving changes in many other genes. Another consequence of genes' involvement in multiple molecular networks is that typically a gene or a functionally significant genetic variation has diverse effects, that is, *pleiotropy*.

The argument as to whether the unit of selection is the gene or the whole organism can be collapsed into a discussion of whether the gene can only be understood in the context of that whole. Some geneticists who have evaluated Dawkins's great idea believe that, while the selfish gene is superficially appealing and correct in particular scenarios, it fails

Immortal. https://doi.org/10.1016/B978-0-323-85692-8.00008-3

in all the ways in which the whole organism, community, or species is the better heuristic framework. The debate as to whether gene, whole organism, community, or species is the level of selection is not entirely a "levels of explanation" conundrum, but a discussion of how genetic variation works.

In this chapter we will see that genetic variation, especially within species, is frequently additive and single locus in effect, as will enable the genetic variant to act as a Dawkinsian selfish gene. Within species, some genetic variants only act *epistatically*, which is to say in combinations with constellations of other genes. As we will see, the importance of epistasis for genetic variants modifying behavior has been vastly overexaggerated. For example, by and large, functional alleles that alter behavior act additively. However, in instances where only the gene combination (i.e., epistasis) will suffice, it is tendentious, or argumentative, to view the single genetic variant as the unit of selection.

The standing of constellations of genetic variations within species is crystallized by the occurrence of long-range linkage disequilibrium in genomic regions containing multiple genes with related functions and evolutionary origins (gene clusters and supergenes) assuring that a cassette of genetic variants will be transmitted together to the same generation. Reproductive isolation followed by speciation is the ultimate assurance of transmitting constellations of genetic variants that must act in combination in order for any of them to increase fitness. For example, a genetic variant that leads to increased wingspan may confer an advantage for soaring, but only if there are concomitant changes in genes encoding muscles, feathers, and behavior. For this reason, it is also equally logical, and mathematically correct, to view gene clusters, with localized combinations of genetic variants, as selfish units of selection. It is also accurate to view a local, reproductively isolated population as a unit of selection. It is true that a species is a unit of selection.

Genes differentially program humans and other species for community, giving animals a variety of tools, both physical—pheromones, color, and such—as well as behavioral—language, mirror neurons, instincts for aggression and cooperativity—for affiliation and reciprocity. Genes determine whether individuals are well-suited to small groups, large groups, or none. Genes determine patterns of social exchange. In many species the tendency to group is turned on and off by biological switches both at different times of the day or year and at different developmental stages. Via inherited genetic variation, natural selection has driven sociocentricity and interdependency of some species to very high levels and made loners of others.

What is the significance of a gene? A main triumph of the genomic revolution is discovery of particulate origins of function and changes in function: genes and variant alleles of genes. However, the question of whether evolution is acting on whole organisms or genes or alleles raises fundamental issues as to what a gene is, and what kinds of evolutionary time frames concern us. The observation that the frequency of a genetic variant has increased across a few generations of selection at the expense of another gene should not make us lose sight of the fact that, over evolutionary time frames, that gene may drive itself, and other genes with which it shares a common genome, into extinction, if the population or species that carries the gene disappears without leaving descendants.

For purposes of illustration and explanation, it is enlightening in several respects to compare genes to words. A word is an ordered collection of letters. A gene is also an ordered collection of the letters A, T, C, and G in the 64 triplet codons, such as TTT, encoding amino acids, in this case phenylalanine, as well as three stop codons that by analogy are periods, which mark the end of genetic sentences, telling the ribosome to stop translating RNA molecules into protein.

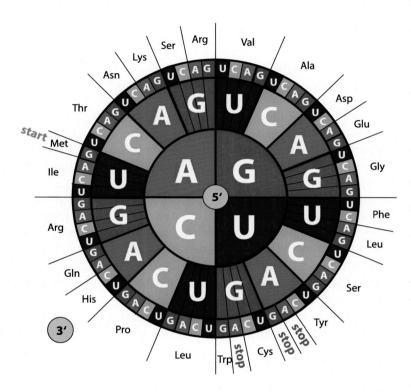

How amino acids are encoded by triplet codons of ribonucleotides in RNA, transcribed from deoxynucleotides in DNA. To read the wheel, the first nucleotide in each triplet codon is at the center of the circle, and the third nucleotide is on the circumference.

Cracking the DNA codes required deciphering the amino acid "words." The first three-letter amino word was identified by Nobel laureate Marshall Nirenberg, who showed that TTT, translated into the RNA UUU, encoded phenylalanine (phe).

DNA words and genes act within the context of a whole language, depending on each other for mutual support, and they coevolve. Superficially it might even seem that the closest genetic analogy to a word would be the nucleotide triplets that are translated into amino acids and stop codons, enabling DNA to first be transcribed into RNA and then translated into protein, much as words are combined into sentences. If so, the genetic lexicon would be very limited, there being only 64 three-letter words formed by A, G, C, and T. However, the codon is not the unit of selection—any amino acid can be encoded into protein sequences of diverse functions much as the same words can form diverse sentences, sometimes by merely changing their order.

Any gene, or DNA element, can be used to make different sentences and stories. Add the word "not" to a sentence, and the meaning is flipped. Most humans have some 3 million DNA sites that are heterozygous, a different nucleotide having been inherited from each parent. In most people, some 50,000 or so of these heterozygous, internally variant loci are located at genes themselves, leading to individuality of DNA encoding proteins.

The most obvious type of gene is an open reading frame with a start codon for initiation of protein translation by the ribosome, and many amino acid "words." However, many RNA molecules have functions other than encoding protein. RNAs can act directly as enzymes, fulfill structural roles, are frequently regulatory, and—when part of protein coding transcripts—can determine the processing, longevity, and translatability of these transcripts. Using the language comparison, some of these other elements might be thought of as elements of punctuation and formatting, and are reminiscent of how the meaning of an entire story or title of a story can be changed by the addition or omission of punctuation: *Murder She Wrote.*

The comparison of bits of genetic material, be they codons encoding a single amino acid or longer sequences of DNA in genes or small regulatory sequences, to words and punctuation is also useful because of how it illumines limitations of the selfish gene model, at least the simplified form of this model that has penetrated contemporary thought.

Words are collected into dictionaries where their different functions (expressions) and to some extent their etymologies (origins) are listed. The evolutionary "success" of a word is crudely demonstrated via the wordle, which quantitates the usage of a word, but also the word can be evaluated for its usage in a language. In what combinations is a word most often used, which may be interpreted as "successful" combinations? Analogously to how genes are assembled into functional networks based on coexpression and physical interactions of their products, the interrelationships between words are identified within grammar and syntax. Linguists following the evolution of words observe that some are highly conserved. Some, like genes, are put to new uses. Words gradually mutate, for example from Old English of Beowulf between the 8th and 11th centuries, to the Middle English of Chaucer in the 14th century, and to the Early Modern English of Shakespeare in the 16th century.

Beowulf, composed by an unknown poet sometime from the 8th to the 11th century:

Hwät! we Gâr-Dena in geâr-dagum
þeód-cyninga þrym gefrunon,
hû þâ äðelingas ellen fremedon.
Oft Scyld Scêfing sceaðena þreátum.

And if death does take me, send the hammered
Mail of my armor to Higlac, return

The inheritance I had from Hrethel, and he
From Wayland. Fate will unwind as it must.

Canterbury Tales (the Knight's Tale), Geoffrey Chaucer, 14th century:

And al above, depynted in a tour,
Saugh I Conquest sittynge in greet honor
With the sharpe swerd over his heed
Hargynge by a soutil twyned threed

And high above, depicted in a tower,
Sat Conquest, robed in majesty and power,
Under a sword that swung above his head,
Sharp-edged and hanging by a subtle thread.

Hamlet, William Shakespeare, 16th century:

Now cracks a noble heart.—Good night, sweet prince,
And flights of angels sing thee to thy rest!—
Why does the drum come hither?

While words have individual independent lineages and destinies, their longevity and fate is mainly determined by the mass of language, tied to culture and circumstance, with which they are coevolving, and other words that may compete with them. Consider the independent selective significance of the Spanish article "las." In Spanish, "las" has a bright and long future as an integral part of a grammar. In English "las" has a more limited future as a component of place names of Spanish origin. Words are at war with each other. Black versus Negro, color versus colour, television versus telly, Baltimore oriole versus northern oriole, and as aptly championed by Gould, *Brontosaurus* versus *Apatasaurus*. The efforts of science bodies such as the French Academy and the American Ornithological Society to pick winners and losers in these word wars largely fail. Lately, the American Ornithological Society has turned its attention to changing the names of birds named after bad people, McCown's longspur now being named something else. The name Audubon still graces birds and ornithological societies, but that could change in an instant and probably without the general public taking note, given that the average person is unaware of longspurs. Mainly, language evolves unpredictably and senselessly, as anyone knows who is "woke." Reason and emotional appeals by readers having failed, there is apparently no way to stop *The New Yorker* from reörnamenting the diaeresis (which happens to be pronounced "die, heiresses"), contriving coöperation from cooperation, preëminent from preeminent, and poëm from poem.

In any culture, new words are relentlessly added and old ones fall from favor, evolve, or disappear. In their internecine competition and cooperation, words work with and against each other in ways surprisingly similar to genes and, as could be discussed, religions or sports teams. Words and genes tend to cooperate and coordinate with others that offer something just a little bit different: for example, the adjective and the noun it modifies, verb and adverb, verb and noun, and words of all kinds that are of a very different feather but that together make communication work. Everyone hated the New England Patriots, due to their sustained excellence and penchant for stretching the rules, and therefore all right-thinking but normally uncooperative fans of other teams cheered whatever other team was playing Bill Belichick, Tom Brady, and the team they led. Like Belichick and Brady, two genes may cooperate because they encode proteins that bind to form a multimeric complex, and again they can work together but at long range by performing very different functions, enabling survival of the body that carries them. Both words and genes compete most remorselessly against the dangerous competitor that is a near version of itself and that executes largely the same function. Imitation may be the most sincere form of flattery, but words, genes, and religions are not ordinarily concerned with flattery. The close but "imperfect" copy is their most likely replacement. In this way, all fans who rooted for the downfall of the Patriots acknowledged their excellence, as well as the evil for which they stood.

As compared to common words, given names mutate even more rapidly, and again their origins and modifications have been traced. For example, the character name that appears most frequently in the works of Shakespeare is "Henry" (seven times, at that). Each word and given name competes with other arrangements of letters for survivability and transmission to the next generation of speakers. The language as a whole, which might be viewed as a mere "vessel" or "robot" transmitting the word, may become a dead language, threatening the survival of most of the words in it.

Before the era of electronic communications and global travel, geographic separation could trigger the rapid divergence of language and the slower but parallel divergence of DNA of people who originated from the same ancestral stock. For example, New Guineans, who are mainly Melanesian peoples, have as many as 800 languages. The European Union recognizes 24 languages, and 40 indigenous languages in Europe are spoken by at least a million people. The world's diversity of languages is a relic of the past, still sustained by literature and tradition but in many cases fast disappearing. As young speakers are integrated into the broader culture of the world, perfectly acceptable languages lose out in the global competition. There are 6000–7000 living languages, and about half have less than 3000 speakers (more than 500 already having less than 100 speakers) and are likely to become extinct in the next generation or two. UNESCO maintains a database and interactive atlas of endangered and extinct languages: http://www.unesco.org/languages-atlas/ complementary to a print edition (Mosely, 2010).

Even when a language dies, it may leave a remnant in languages that survive it. Within languages, certain words, predicted by their high frequency of use, are more long-lasting, and their evolutionary trajectories within and across languages are used by linguists to trace the affinities of languages across millennia. Because of the extreme antiquity of the two dozen words that comprise them, these four sentences might have been understood 15,000 years ago by hunter-gatherers in Asia, as they are today by English speakers (Pagel et al., 2013):

> *You, hear me! Give this fire to that old man. Pull the black worm off the bark and give it to the mother. And no spitting in the ashes!*

These words, highly conserved though they are, are a pitiful remnant of the past as compared to amino acid triplet codons, and whole selfish genes that have survived hundreds of millions of years essentially unchanged, and that were essential to species long extinct. However, when languages or groups of languages die out, thousands of words and even the mainstays are lost or become archaic, existing only as fossils, perhaps to be deciphered from a stone engraved in a lost city, for example the ancient languages of the Maya and Egyptian hieroglyphics that were decrypted centuries after their rediscovery.

Given that words make sense in the context of grammar, one may properly rebel at defining a word as a particle of a language. However, it is more difficult to define "gene." In the decades since Dawkins wrote *The Selfish Gene*, it has become appreciated that most of our genomic DNA, probably more than 90% of it, is at least occasionally transcribed into RNA and that even the most extensively transcribed regions make diverse RNA products that are subsequently processed into many forms. Expression is insufficient to define function, although this was the standard adopted by an international genomics consortium, ENCODE (Encyclopedia of DNA Elements), devoted to unraveling the function of the genome. However, many "nongenic" regions of DNA are functional, while other occasionally transcribed sequences are not, and tend not to be conserved evolutionarily. In defining genes, the answer is not "either/or." Nontranscribed DNA can itself be functionally critical, for example altering DNA structure to increase or decrease transcription of DNA, serving as a recognition sequence for DNA recombination during meiosis, regulating nonmeiotic recombination as happens at human immunoglobulin genes and leading to the incredible diversity of antibodies, and forming the centromeres and telomeres that are integral to the structure and function of chromosomes.

ENCODE, and earlier individual groups of investigators, observed that most of the human genome is transcribed into RNA and that thousands of these indeed represent alternate protein encoding forms and small RNA regulatory transcripts. However, other transcripts are found at only low levels and at less than one copy per cell on average, even in the cells that express them. The apparatus controlling DNA transcription is imperfect, in something of the way that an apparatus recording and repeating speech is fallible. Suddenly, and perhaps to confuse a linguist or copyeditor, a random assemblage of letters might appear on a page, "Kiopy ujytef mmigyf!!" And never to be seen again. Brilliantly, Charles Dodgson exploited this tove.

Nevertheless, the discovery of widespread transcription of DNA into RNA, the examples of nonprotein coding RNAs that are regulatory, and ENCODE's definition led to a popularization of the idea that 90% of the genome is functional. At the present time, we are aware of some 50 million base pairs of the human genome that are highly and often expressed. This so-called exome includes some 2000 regulatory microRNAs and tens of thousands of regulatory small RNAs, as well as the structural RNAs that comprise the ribosomes and nucleolus and that perform the enzymatic function of splicing other RNAs. Based on the ENCODE data, and its tautological definition of transcript functionality, people from other disciplines such as engineering and systems design argued that the vast portion of the genome exists not to harbor 25,000 protein coding genes but to encode hundreds of millions of regulatory RNAs. As usual, it was asserted that humans were thereby special, because our genome is a bit more complex than some other primate species whose complement of genic sequences and DNA sequence identity are otherwise so annoyingly similar to ours—a chimpanzee has almost every gene a human has, and is 99% identical in sequence. So might the modest expansion of genome size as compared to those particular species have happened because of the greater regulatory complexity needed to construct the human brain?

Probably not. Certain other vertebrate and even single-celled species have dramatically larger, and smaller, genome sizes than *Homo sapiens*. At 133 megabases, the genome of the leopard lungfish, *Protopterus aethiopicus*, is 40 times larger than the human. At 670 megabases, the genome of *Polychaos dubia*, a single-celled amoeboid creature that lives in fresh water, wins the prize for size. It is doubtful that the lungfish is 40 times as complex as a human or that an amoeba is four times as complex as a lungfish. When the transcriptomes of *Protopterus* and *Chaos* are analyzed, it would be unsurprising if 90% of their giant genomes are, like that of the human, also transcribed occasionally. Whole genome functionality based upon rare transcripts is at best an unproven idea, having not been supported by sufficient observations of roles of the rarely transcribed RNAs or their evolutionary conservation.

Meanwhile, added layers of DNA, RNA, and protein epigenetic regulation are providing a practically infinite variety of ways for a cell to regulate molecular functions, without invoking functionality for all of the larger component of the DNA sequence that is evolutionarily nonconserved.

Even taking this conservative view of genomic functionality, the variety, number, and overlaps of functionalities in the genome still largely defeat the gene conception. Usually, expressed regions of the genome ("genes") have multiple transcription start sites and can be differentially processed in a variety of ways. Depending on how this is done, a different functionality is expressed. In numerous regions of the genome multiple sequences with interregulated functions are colocated within only tens of thousands of nucleotides of each other, so that they can be coregulated (when one is turned on, the other is turned on) or counterregulated (when one is turned on, the other is turned off) or as artifacts of their origins as ancient gene duplications. Sometimes one gene is located in a region corresponding to the intron (spliced out) region of the other RNA transcript. Not infrequently, two different proteins are encoded by the same stretch of genomic DNA, the protein encoded depending on whether the DNA strand or the complementary DNA strand is transcribed. Also, even if a particular functionality of a sequence of DNA has been highly conserved, there is likely to be an accumulation of many sequence changes over evolutionary time frames and most of these changes do not alter function or fitness. In part, this is due to the degeneracy of the DNA code as it encodes amino acids. As seen in the figure, there are 64 triplet DNA nucleotide combinations that only need to encode 21 amino acids and the stop codon. Therefore, many mutations are synonymous, which is to say that they do not alter the amino acid sequence of the protein in some open reading frame that is not disrupted by a stop codon.

In other words, as we search earnestly for the "selfish gene," we come up against the problem that a built-in feature of the genetic code guarantees that few genes will be successful in faithfully transmitting themselves over evolutionary time frames, and indeed few genes have been. Across species, the rate of neutral substitution is approximately 1 in 100 nucleotides per million years, with the result that even in a so-called living fossil such as the coelacanth or tuatara, most genes differ in sequence, and are subtly different functionally, from the version that existed 100 million years ago.

Another challenge to gene selfishness is the enormous amount of variation within populations, even within the human population. Humans are not as variable as many other species because of our relatively low effective population size: an effective population size of only 10,000 or so who might have freely mated, in the recent past. As elucidated by Masatoshi Nei, the major factor that shapes the level of genetic variation is neutral genetic variation whose level is determined by effective population size and neutral drift rather than genetic selection and selective sweeps in the DNA regions that are nearby functional loci. In other words, DNA variation – and even the variation within the functional bits of the genome that have been labeled "genes"—is shaped more by random processes than by Darwinian selection. There were many more than 10,000 humans throughout most of the evolution of *Homo sapiens*. Nevertheless, people were less mobile. Bounded by locality and the accidents of geography, they were unlikely to mate with someone born in the next valley or across an ocean. Even populations in proximity might not admix because of language, custom, or enmity. This locality of mating, and relatively small effective population size, has been verified by researchers who asked how likely it was that a person in the first part of the 20th century would mate with a first or second cousin. Such studies, conducted in cultural settings where first and second cousin marriages were not taboo, found that cousin matings occurred far more often than would be expected by chance and with a frequency compatible with an effective population size of 10,000, give or take a few thousand.

Over evolutionary time frames, the random nucleotide substitutions in DNA and amino acid substitutions in proteins tick at a fairly constant rate, determined by mutation rate of the species. As such, the nucleotide substitution rate is a molecular clock, as originally postulated by Linus Pauling and Émile Zuckerkandl in the 1950s. Vincent Sarich and Allan Wilson used protein analyses based on antibodies raised against common proteins to test the molecular clock hypothesis. If two species' albumin proteins were more divergent, antibodies raised in these species against the other species albumen would cross-react more strongly. Sarich and Wilson not only proved the molecular clock hypothesis but upset ideas about mammalian phylogeny that were based on the fossil evidence then available. The protein evidence showed that humans had diverged from other hominoid apes only within the past 10 million years (and eventually it was narrowed to within 5 million), rather than 20 million years ago.

Thus far, some 22 million common genetic variants have been found in people living today, and as individuals we are all genetically variable at approximately 1 in 1000 nucleotides, at which we have inherited a different allelic form from our mother than from our father. This rate of heterozygosity translates to about 3 million genetic variants in each person. Although most of these are selectively neutral, many are decidedly nontrivial. For example, people typically are carrying more than 10 stop codons capable of blocking the function of a "gene" transcript and dozens of large copy number variations, some of which duplicate or delete entire "genes." Selection happens, and there are many specific examples in which mechanism and effect have been shown, and a few are discussed elsewhere in this book. However, on an overall basis neutral drift and other processes more often defeat selection when it comes to just which genetic "words" will be found within people and transmitted to succeeding generations. If genes were more effectively selfish, one would think they would have built a better mechanism to preserve their integrity and a more consistent result.

On the other hand, the lack of conservation of genes may point to the conservation, and selfishness, of higher-order constructs, for example the species. Reminding us of homeostasis, some living species appear to have been conserved across hundreds of millions of years, during which the molecular (genic) clock was constantly ticking. The coelacanth, a so-called "living fossil," was discovered off the coast of South Africa in 1938, and although unappetizing is occasionally caught as bycatch by fishermen trawling in deep waters. The Comoros coelacanth (*Latimeria chalumnae*) and Indonesian coelacanth (*Latimeria menadoensis*) are the only extant species of *Latimeria*, an ancient order of lobe-finned fishes from which lungfish, mammals, and humans evolved, and which was thought to be extinct. We cannot know many of the aspects of the physiology of ancient coelacanths, but modern coelacanths appear remarkably similar, if not identical, to the coelacanth in fossils that are 400 million years old. Furthermore, the coelacanth has many bizarre features that distinguish it from other fish. Instead of a vertebral column, it has a notochord, a hollow pressurized tube filled with oil. The braincase is almost entirely filled with fat. It has a unique vestigial lung filled with fat. The kidneys are displaced ventrally in the abdominal cavity by this weird organ, and fused in their location. What seems most conserved in coelacanths is not individual genes or organs, but a constellation of physiology, organs, and genes enabling it to control buoyancy and survive in benthic habitats. The "platonic ideal" of the coelacanth is conserved even if the genes and parts they encode have undergone substitutions, and minor improvements here and there, like a vintage car or historic house that gets repainted and reshingled.

A fossil coelacanth.

In search of the successful selfish gene, there appear to be only a few hopeful candidates. The leader might be the Ribonuclease P gene that encodes an enzymatically active RNA working within the cell's ribosomes to translate RNA to protein. As described elsewhere in this book, an RNA something like Ribonuclease P could have been the replicator molecule ancestral to all life on Earth, and tRNAs mediating the highly conserved translation of the DNA code to specific amino acids are a close second.

In the category of DNA-related genes, histones are a group of extraordinarily successful and conserved genes whose function is right at the heart of the workings of DNA. If stretched out, human DNA would be several feet in length and one can imagine that the chromosomes soon would become hopelessly tangled like Nadia's necklaces in a jewelery box. Instead it is packaged and organized into a DNA/protein complex called chromatin, the most important component of which are the histone proteins. There are multiple highly conserved histone proteins of ancient origin and there are numerous molecular processes in the nucleus of our cells that have evolved to recognize and modify histone proteins, for example by adding and cleaving acetyl (two carbon) and methyl (one carbon) groups at specific sites on the histones and thereby altering chromatin structure. A major achievement of the ENCODE consortium discussed earlier has been to use the pattern of histone composition to identify several types of chromatin structure that are in different regulatory states.

One hopeful way back to the selfish gene would be via evolutionary conservation. However, as is now appreciated, in the end it appears to be simplistic to accept any individual stretch of DNA nucleotides as the unit of selection, although it may be convenient or even necessary to do so in some circumstances.

It is worth thinking about cross-species competition from the perspective of an extraordinarily conserved selfish gene. The histones are extremely ancient and evolutionarily highly conserved proteins. We essentially share the same histones as yeast. Therefore, from the standpoint of a histone, a very effective way of making copies of itself might be to encourage, in cross-species fashion, the cultivation of yeast, for example by brewing beer. Clearly, there are limits to the selfishness of a gene, because until quite recently, when histone genes of humans and yeast were sequenced, no histone gene had any awareness of its identity to the gene in another species, or its own. Indeed, a histone gene does not have identity or awareness, as such. However, if yeast and human histone genes could, while unaware, reciprocally facilitate the others' transmission, they would.

An example of a very selfish gene is the transposon, or "jumping gene." Barbara McClintock, at the Cold Spring Harbor Laboratory, discovered transposons by observing the hypermutability of genes that could alter the color of individual kernels in maize, leading to a varied and multicolored ear of corn. Later *Drosophila*, which has different types of transposons and in which transposons are now used as tools for genetic manipulation, provided one of the finest examples of the mechanism and evolutionary significance of transposons when a particular transposon known as the P element in essence made a new species. *Drosophila* geneticists suddenly became aware that mating between laboratory strains of *Drosophila* and wild flies had become dysgenic. The hybrids showed a dramatic reduction in fertility, with their surviving offspring manifesting all kinds of mutations, and this occurred because the transposase activity of the P element was suddenly de-repressed in the hybrids. The transposon thus can be a factor encouraging reproductive isolation and speciation, thereby conserving the relationships between constellations of genes that are found in the new species.

A transposon is something like a computer virus. A complete transposon is a relatively small genetic unit that lives a parasitic existence in the genomes of many or most species. There are many types of transposons and several have altered large swaths of the human genome. The human genome shows signs of heavy infection. Approximately 40% of the human genome consists of mostly dead, no-longer-active transposon sequences, mostly the partial sequences of the LINE (Long Interspersed Nuclear Element) and SINE (Short Interspersed Nuclear Element) families of transposons, the LINEs being about 7000 DNA bases long and the SINEs being nonautonomous, on-coding elements anywhere from 100 to 700 bases in length. This is not to say that transposons do not have beneficial effects, for example in promoting speciation via hybrid dysgenesis, but 40% of the genome is much genomic space given over to useless and potentially damaging elements that occasionally transpose into genes and other needed genetic elements and thereby disrupt them. In humans, some 4000 of about 100,000 LINEs we carry are full-length, and although their activity is actively repressed by epigenetic silencing, occasionally they transpose, leading to a variety of inherited genetic diseases and cancers, thew latter if a proto-oncogene such as a gene that regulates cell cycle happens to be abnormally activated or deactivated. These sequences were deposited in our genome recently in evolutionary time and are not even found in our closest great ape relatives.

A complete transposon such as a full-length LINE consists of a palindromic recognition sequence at each end with a sequence encoding the transposase enzyme sequence in the middle. When the transposase sequence is transcribed into RNA and translated into protein, it becomes a transposase enzyme that will recognize itself or other transposon sequences, including incomplete ones, elsewhere in the genome. It can cut them out, copy them, and splice them into new genomic locations without regard to the needs of the host genome. If the transposon happens to disrupt an essential DNA sequence, the results can be calamitous. Occasionally a mutation induced by a transposon would even be beneficial; however, this happens so rarely that cells do the best they can to suppress transposons.

In addition to neutral genetic variants, each person also carries genetic variants that are functional, and that thereby have consequences for physiology and physiognomy. Some of these are rare functional variants, for example the rare variants of *HPRT* (Hypoxanthine Guanine Phosphoribosyltransferase) that cause Lesch-Nyhan syndrome, that are being actively selected against and occasionally replaced by new mutations. These are uncommon in any population, having a short half-life.

However, there are also functional variants that have survived for hundreds, and even thousands, of generations and that are common in populations worldwide. At such genes, two functional alleles are maintained by balanced selection, neither having a decisive advantage over the other. For some, there is niche dependent selection, heterozygous advantage, frequency-dependent selection, or time-dependent selection.

Within the human species, some genetic variants that are functionally variant, and constellations of them, have at least a limited ability to preserve themselves via assortative mating. At the top of the list would be genes that encode features that are directly selected for in mating. In this context, many species have mating rituals, for example elaborate dances and mating displays, contests of strength, and nest building, that may enable selection for constellations of genetic variants. The animal with a genetic variant leading to facial asymmetry, altered coat color, lack of vigor, or inability to hold a territory against competitors will tend not to transmit that genetic variant. Assortative mating for some multigenic human traits, for example general cognitive ability (sometimes called IQ) and height, and addictions, is profound. This assortative mating can, to a small extent, bind together constellations of genes, but the reproductive isolation conferred by ordinary human assortative mating is ineffective if a large constellation of genes is required to epistatically produce a phenotype. A phenotype requiring the interaction of 10 genes would be seen once, but probably never again without reproductive isolation, as might happen due to transposon-induced infertility of isolation of populations. In people, the use of race or of other constructs such as "liberal" or "conservative" are weak and ineffective means of preserving such gene constellations. Yet, it is obvious that people use them.

Race: Most of these "selfish genes" would be unable to preserve and propagate themselves using racial identifiers, but this has not stopped people from trying, for example, to preserve the "Aryan race." This is certainly true for common genetic variants such as exist in behaviorally significant genes such as *COMT, NPY, BDNF, SLC6A4,* and *MAOA.* All of these genetic variants alter behavior, but whereas a person might assortatively mate with another based on behavioral similarity, their contributions to behavior, be it anxiety or executive cognitive function, are too small to make this assortative mating very effective, at least without the help of a geneticist. However, in the first half of the 20th century, genetics, and the eugenics movement composed of leaders in genetics, provided a further false justification. As Gertrude Stein remarked, if something cannot go on forever, it will stop. However, the ends of racism, the consequences of past racism, are not in sight. If people did not allow race, culture, geography, and economics to determine their choice of mate, interracial marriages would be much more frequent.

Assortative mating (*homogamy*) is easily observable in the mating patterns in the United States. Elsewhere, the occurrence and importance of *endogamy*—the tendency of blood relatives to mate—is discussed. Proving that homogamy is the rule rather than the exception, in 2014 according to the United States Census, the frequencies of four so-called races were White 77%, Black or African American 13%, Asian 5%, and Hispanic or Latino 17%. All other things being equal, the odds that a male or female of one race would marry someone of the same race were 77%, 13%, 5%, and 17% for the four groups, respectively. Instead, people were very likely to mate with people of the same race, whether because of race itself or because factors that powerfully cross-correlate with race were causing them to assort by race. In 2013, interracial marriages accounted for only 12% of newlyweds in the United States (according to a Pew Research poll, APS Integrated Public Use Sample, 2013, http://www.pewresearch.org/fact-tank/2015/06/12/interracial-marriage-who-is-marrying-out/). Only 19% of Black Americans (instead of 87%), and only 28% of Asian Americans (instead of 95%) married someone of a different race. Black women had only a 12% chance of marrying outside their race, representing a five-fold assortative mating effect.

For complex traits, assortative mating is only moderately effective in preserving constellations of genetic variants that affect those traits. Elsewhere, the importance of large chromosome segments (*supergenes*) containing constellations of genes and the role of the sex chromosomes in harboring gene constellations altering behavior, is discussed. Children's IQ and height correlates more highly with the mid-parent average than to the IQ or height of the individual parent. However, assortative mating performed without the help of a DNA sequencer is *relatively* ineffective as a means of preserving the genetic combinations that count. Obviously, assortative mating can be fairly effective for a trait such as height. For IQ, the spousal correlation is only about 0.33 (Bouchard and McGue, 1981).

The assortative mating, to the extent it works, is effective because some of the genetically influenced behavioral traits are multigenic, with each genetic variant inherited from the parent polygenically adding to the overall, quantitative trait, such as IQ. To a lesser extent than for IQ, people assortatively mate for personality, as may be enhanced by modern dating websites. In so-called unstable marriages, there may be dis-assortative mating for some personality traits, for example outgoingness (Cattell and Nesselroade, 1967). However, race, the factor against which people are most likely to mate assortatively, has little to do with intelligence or personality. The racial categories are amalgams of many different populations. For example, Hispanics include Europeans of Hispanic descent and Hispanic populations with widely varying degrees of European, Indian, and African ancestry. Africans are the most genetically diverse of the world's populations.

Ancestry scarcely predicts human nature, but people, perhaps under the influence of ancient kinship drives, are fascinated by their ancestral origins. As of the time of this book, some 1.3 million people had used the services of 23andMe to learn more of their ancestry, and other companies offering such services were also flourishing. By measuring ancestry with genetic markers, it can readily be determined that the actual African ancestry of Black people in the United States ranges widely, from close to 100% to close to 0%, although all humans can trace their ancestry to Africa only a little more than 100,000 years ago.

The genomic revolution, rather than leading people to treat each other as individuals, has stimulated in many a fascination with their origins, who they are more than what they are. Although it is increasingly becoming possible to use genes to define what we are, we still know too little about the genes influencing brain and behavior to make powerful predictions, but with some certainty we can estimate components of ancestry. Genetic testing offers a new means by which people can identify others with whom they share closest genetic affinities, excluding others. Via genetic identifiers, whether the identifier be an externally identifiable marker such as skin color or a panel of DNA markers, people attempt to recapitulate the kin identification essential to altruistic behaviors intrinsic to inclusive fitness.

The significance of much of the information provided on ancestry, whether from pedigree charts or from genes, is overinflated. For example, and as was the premise of *The Da Vinci Code*, if a person learned that they were the direct descendent of Christ, and separated by about 100 generations, genetic relationship to that Messiah would be $(1/2)^{100}$. Effectively, zero. It seems a harmless diversion to find evanescent connections to the past, but it can easily lead to misunderstanding and discrimination in the present. The woman who thought she was of German ancestry, and perhaps who did not care much one way or another, suddenly is told she is Irish and begins adopting accoutrements of Irish identity.

Reminding us of the fascination of Germans with the genetics of the *Volk*, Chinese geneticists are attempting to bring their ancient cultural legends to life. At Fudan University's Key Laboratory of Contemporary Anthropology, some 400,000 DNA samples have been collected from all over China because, as a leader of the study said, "Chinese people want to know their links to each other." "Before DNA, they liked to study their family records. They want to know who they are related to" (Kathleen McLaughlin, Science, 2016). The laboratory appears to be making excellent headway in its nationalist project. A signal discovery was their finding that Taiwanese are an integral part of the Chinese *Volk*, and not descendants of Polynesians. This buttresses China's insistence that Taiwan is a province of China and not an independent nation.

The genetic anthropology at Fudan also plays a role in internal identity politics of China. It is commingled with criminal forensics by the apparatus of the state. Using the DNA profiles in the database, collected from volunteers, not criminals, the Key Lab's geneticists have identified criminal perpetrators. The accused does not have to be in the database because he can be identified by genetic resemblance to someone who is in the database, and much more so than in many other countries. For centuries the Chinese have been relatively immobile, during which time most clans of genetically related people were restricted to a home village or local region. A suspect's surname is also diagnostic because most Chinese share only 100 family names. For example, serial killings in Gansu province were traced to a man from that region with the family name of Gao. Although Gao had never been directly genotyped himself, it had been deduced that the killer was a Gao, from Gansu. Gao confessed. A main ongoing project of the Fudan Key Lab is to prove the existence of the Three Sovereigns, three ancient, mystical, figures in the pantheon of China's past. Indeed, about half of Chinese men fall into three ancient Y chromosome groups, which convinces the leader of the study that he is on the right track: three ancient Y chromosomes, three sovereigns, one Chinese racial identity dating back at least 6500 years. This is genetics with a purpose.

Hostility and discrimination directed against the stranger is ancient, and ingrained throughout human history and prehistory. Due to kinship selection, we evolved to be kind to blood relatives and less kind to those to whom we are unrelated. When a population of unrelated individuals can be identified, it has frequently been singled out for maltreatment. The worst victimized, whether identifiable by culture, language, or origin, have also often nursed grudges over years or even centuries, and then turned the tables on their oppressors, their descendants, or bystanders, becoming perpetrators themselves. The cycle of hatred never ends in parts of the world where people demand expiation or payback for thousand-year-old grievances. The implication that payback is necessary also encourages the denial of historical outrages. People refuse to squarely acknowledge the travesties of their ancestors because they feel that somehow they are morally responsible for what someone did 100 years before they were born, and in denying, fail in their own moral responsibility.

Clearly, cultural memory plays a role that often overwhelms genetic predisposition to racism. Racism provokes racism, and hate provokes hate. Racism based on superficialities such as skin color failed, although the assortative mating continued. However, with molecular genetics, there is a new way to discriminate against people on a genetic basis. For example, we could easily specify that an organ be donated to a distant relative identified by a genetic profile in a large database. Many people using personal genotyping services take advantage of the option to post their results, so that others who are their distant kin can identify them and themselves, reconstructing in our modern world the sort of deep kinship structures that tribal peoples relied upon to determine with whom they would ally or have conflict. The result may be good: for example, more organs donated. However, the effect can also be to fracture society along lines of genetic ancestry, based on a false premise of affinity and similarity in the deep aspects of intellect, personality, and talent that can be called humanity.

Racism is not based in the science of genetics, having preceded any discoveries of genes or heritability of individual characteristics, and having been refuted by the discovery that human variation is more than 90% interindividual and only to a small extent population-based. The attempts of racists to use physiognomy to identify and denigrate the outsider are well known, for example, the classically racist Nazi characterization of the appearance of Jews. However, except for the use of skin color and other obvious markers such as language and dress, racism has been failing. Via DNA sequencing of thousands of people, and very large-scale genotyping of tens of thousands, we have learned that skin color and other surface physical characteristics, language, and ancestral origin are very poor predictors of the totality of an individual's genetic variation. Racism differs from an understanding of physical differences, cultural differences, religious differences, and genetic differences. It is a self-amplifying belief system justifying the mistreatment of a group and we will discuss how these belief systems take advantage of minute physical, genetic, and cultural differences, accentuating differences rather than reaching for commonalities.

In this chapter we have seen that the consequences of taking a biological specimen from a living or dead person are potentially far-reaching because of the DNA code that is contained therein. Using that code, one may make all kinds of inferences and discoveries about the person and their relatives and community, both living and dead. A gene from an ancient, extinct species can be as active and potent as a word from a language that is otherwise long forgotten. Some of the gene sequences are themselves incredibly successful and ancient selfish genes that themselves fit the definition of a living entity within our genomes. Carried within the skeletons of the dead are the DNA sequences that encoded the shape and behavior of the whole person. These DNA words and paragraphs are not by any means the whole story of the person—which would diminish the individuality of twins and clones. However, the DNA "message in a bottle" is a profound one, with implications to which a person may not have wished to consent, and future implications that are difficult to anticipate.

9

The ancient divide between molecule and self

Is moral value conveyed to the ingredients, molecules and cells, of which we are made? Beyond the stories told in a person's DNA, it is an argument of this book that a cell or molecule can have value because of provenance. This is not to assert equivalence. We respect the persons from whom they were taken, regardless of whether that person is dead. However, this chapter asks the question as to whether certain molecules and cells have a further independent value if they represent life-stuff—something on our side of the boundary between death and life, intrinsic to human origins. If a thing is ancestral to oneself—a progenitor and not merely an ingredient—it is part of a living continuum of which humans are only the most recent manifestation.

This evolutionary argument for the value of things is nonequivalent to an assertion of the sacredness of all life, or to the moral worth of food that fuels the body, and that is used for day to day replacement of many of the atoms in the body. With each breath a person inspires approximately 18 mg of oxygen, some 6×10^{20} molecules, and expires approximately 6×10^{20} molecules of carbon dioxide and approximately 6×10^{20} molecules of water. It would be preposterous to try to concern ourselves with the fate of each molecule exchanged in one breath, except in the mystical sense that each person is linked in countless ways to all things on Earth, both in substance and in binding networks of causality. Mystics recognize the connection between all things, and magicians believe these links can be manipulated. Scientists, engineers, and others of their ilk trace physical interactions, learning that some interactions are primordial, and some manipulate the connections between things—and are not necessarily the same ones. Tracing the ancestry of humans, it has been learned that some cells and molecules were more than ingredients of life—some were the primordial ancestors from which humans and other life on earth descended.

The origin of cells, and from cells, consciousness, is the story of life on Earth that unfolded across 3 billion years of evolutionary tinkering fueled by mutation and guided by natural selection. From chemistry (molecules) to cells was relatively fast—a space of tens of millions of years. The transition from cell to multicellular (metazoan) life was slow, requiring a billion years. Cells had a lot to learn, and conditions had to be suitable for the evolution of multicellular life. However, once multicellular life evolved, things happened "quickly," which is to across epochs of only hundreds of millions of years. Once primates evolved some 55 million years ago, it seems to have been evolutionarily certain that a species with the cognitive capacities of *Homo sapiens* would evolve, because we evolved very rapidly indeed, and within a space of only 20 million years from an ancestral ape.

Immortal. https://doi.org/10.1016/B978-0-323-85692-8.00009-5

Charles Darwin's and Alfred Wallace's papers on evolution were read before the Linnean Society of London in 1858. The theory of evolution, which is actually a fact encompassing vast complexities and several acute controversies, was thus conceived only in the last instant of evolution of life on earth. Today, belief in evolution is pervasive among the educated, permeating scientific, philosophical, and social discourse, but the theory's significance was not initially appreciated. Reviewing the Linnean Society's activities in 1858, its president Thomas Bell wrote, "The year which has passed … has not, indeed, been marked by any of those striking discoveries which at once revolutionize, so to speak, the department of science on which they bear."

Brains, cells, and DNA are the stuff and consequence of evolution. All life on Earth discovered so far is related, however distantly its forms have diverged. This chapter is about the primordial relationship of people to other organisms, cells, and molecules. It is about the boundary between life and chemistry, focusing on the difficult, and still mysterious, early steps in the process, as well as how multicellular life first appeared. In the beginning, on Earth or elsewhere, there was a self-replicating molecule. How can a cell, with 25,000 or so genes, interlocking molecular pathways, and complex physiology, have evolved from a single, ancestral, self-replicating molecule? How, in a universe where entropy relentlessly increases, can cellular, and ultimately neurocognitive, order have arisen out of chaos?

The misconception that complex life arose by random mutation is popular, perhaps because this inaccuracy serves a "higher" purpose. For example, tracts from Jehovah's Witnesses, kindly given to the author by his erstwhile proselytizing friends "Peter" and "Joseph," lay out uncertainties about the origin of life and consciousness that trouble many. How could a work as marvelous as humankind have come into existence except by the hand of a creator? It could not possibly have happened by random DNA dice rolls. This vital issue has been explored, both clearly, and early, in Dawkins' *The Blind Watchmaker*. Something about the message remains elusive, perhaps because the explanation has been so simplified that deeper points have been elided. The purpose here is not to quash creation science. It is a step forward that people interested in human nature turn to scientific evidence, even when that evidence is misunderstood, misinterpreted, or misrepresented.

In linking humans to all other life on Earth, it would be inconvenient if not fatal to the project if life was purposefully designed or synthesized *de novo* from elements at hand, as in the Book of Genesis. The link between life-forms would thereby be God rather than shared ancestral roots. People would be related to paramecia or piranhas only in the same way that different models of cars designed and manufactured by General Motors are related. Whether evolution is intelligently directed, or designed, is a separate conversation that does not have the same negative bearing on the interrelationship of life. In a sense, evolution was "guided" by an external hand via the pressure of natural selection. However, there is no evidence of intelligent design. If evolution was intelligently guided, the guidance was performed with subtlety and deception, because nothing about evolution appears intelligent beyond the predictable effects of selection, and against a background of unintelligently enormous wastage.

How can life have evolved via the random accumulation of mutations? Well, it did not. This would be less realistic than asking a million fun-loving monkeys to randomly recreate *Hamlet*. The laws of thermodynamics describe relations between three fundamental physical properties of the universe: temperature, energy, and entropy. Of these properties, *entropy*, describing the disorder of a system, is seemingly at odds with life. By the second law of thermodynamics, formulated by Sadi Carnot in 1824, *entropy* (disorder) of a closed system always increases. How can living systems be so highly ordered, and nonrandom in nature, and how can their forms be encoded and faithfully instantiated into molecules, physiology, and cells? By the mid-20th century, physicists such as Max Delbrück turned their attention to biology in hopes of learning new principles of physics. New physical principles did not emerge—life obeys the same rules that govern the behavior of other matter, but life is more complicated. However, they helped launch the science of molecular biology, which led to the discovery of the DNA code and mechanisms by which cellular functions unfold. With the identification of DNA as an information molecule capable of being shaped by mutation and natural selection, Darwin's insight that natural selection drove evolutionary change makes sense of the origin of complexity in living organisms and their constituent cells, all of which are decidedly ordered and nonrandom in nature. Life cannot reverse the entropic arrow of time, but living beings are eddies in that current, being islands of locally reduced entropy created at the expense of an overall increase in disorder. The answer to the riddle of life is the guiding hand of natural selection, Dawkins's "Blind Watchmaker."

The 19th century was a revolutionary time in biology when most of the fundamental concepts that guide biologists today were discovered. Also in the 1800s, Ernst Haeckel realized that ontogeny (development) recapitulates phylogeny (evolution). It is important to grasp the limitations of Haeckel's statement, for example the imaginal discs that form the adult organs within an insect larva do not recapitulate phylogeny in any exact way. However, there are many surprising parallels in the way that biological order and complexity have arisen from simple, chaotic systems both evolutionarily, leading to multicellular organisms, and in the development of the body, which also begins as a single cell and recapitulates, in a sense, the evolutionary process.

Alfred Russel Wallace, the codiscoverer of evolution and its main driver natural selection, believed that divine intercession was necessary to explain three great events in the history of life: the origin of life from inorganic matter, the genesis of consciousness in higher animals, and the origin of higher mental faculties such as music and mathematics in humankind; he also believed that the universe existed to enable the development of the human spirit. Wallace's ideas that evolution, and the universe's existence, were teleologically and anthropocentrically driven anticipated modern creation science, as pointed out by science historian Michael Shermer. In the frontispiece of *The World of Life*, Wallace quotes Alphonse de Candolle: "Every plant, whether beech, lily, or seaweed, has its origin in a cell, which does not contain the ulterior product, but which is endowed with, or accompanied, by a force which directs the formation of all later developments. Here is the fact, or rather the mystery, as to the production of the several species with their special organs." Wallace himself wrote (p. 277, *World of Life*), "…beyond all the phenomena of nature and their immediate causes and laws there is Mind and Purpose; and that the ultimate purpose is (so far as we can discern) the development of mankind for an enduring spiritual existence." "If this view is the true one, we may look upon our universe, in all its parts and during its whole existence, as slowly but surely marching onwards to a predestined end." Wallace passionately advocated reverence for life based on his understanding of creation:

> To pollute a spring or a river, to exterminate a bird or beast, should be treated as moral offences and as social crimes; while all who profess religion or sincerely believe in the Deity – the designer and maker of this world and every living thing – should, one would have thought, have placed this among the first of their forbidden sins, since to deface or destroy that which has been brought into existence for the use and enjoyment, the education and elevation of the human race, is a direct denial of the wisdom and goodness of the Creator, about which they so loudly and persistently prate and preach.

Foreshadowing the modern debate about neurogenetic determinism, Wallace was in part responding to his contemporary Haeckel who had written in *The Riddle of the Universe*, "The peculiar phenomenon of consciousness…is a physiological problem, and as such, must be reduced to the phenomena of physics and chemistry." Haeckel originated many terms in biology including phylum and ecology, and among his inventions was a <u>monistic</u> philosophy linking science and religion, casting evolution as a cosmic force. Haeckel's monism was not easy to pin down, and evolved over his lifetime, but stimulated materialistic (atheistic) theories of origins such as this one that Haeckel quoted, "The development of the universe is a monistic mechanical process, in which we discover no aim or purpose whatever; what we call design in the organic world is a special result of biologic agencies; neither in the heavenly bodies nor in the crust of the earth do we find a trace of a controlling agency – all is the result of chance."

Especially distasteful to Wallace was this statement of Haeckel's:

> Our own 'human nature' which exalted itself into an image of God in an anthropistic illusion, sinks to the level of a placental mammal, which has no more value for the universe at large than the ant, the fly, the microscopic infusorium, or the smallest bacillus. Humanity is but a transitory phase of the evolution of an eternal substance, a particular phenomenal form of matter and energy, the true proportion of which we soon perceive when we set it on the background of infinite space and eternal time. (p. 87)

JBS Haldane took down creationism before the ideas of intelligent design and creationism became popular. Asked what evolutionary biology disclosed about the mind of God, Haldane tartly answered that God must have been inordinately fond of beetles, having created so many:

> *The Creator would appear as endowed with a passion for stars, on the one hand, and for beetles on the other, for the simple reason that there are nearly 300,000 species of beetle known, and perhaps more, as compared with somewhat less than 9,000 species of birds and a little over 10,000 species of mammals. Beetles are actually more numerous than the species of any other insect order. That kind of thing is characteristic of nature.*
>
> What is Life? The Layman's View of Nature, p. 248, 1949.

More damaging to creationism, evolutionary biologists after Haldane fleshed out the details of catastrophic, and often random, failures of countless species that had been hinted at by the catastrophism observed by Cuvier. Evolution works for eons via chance (mutation) and necessity (adaptation) to perfect a life-form only for a random meteor to destroy it in one day. The life form is then never seen again, its niche having been occupied by another species, and the life stuff from which it evolved having long since disappeared. Evolutionary missteps, dead ends, catastrophes, and mass extinctions are plentiful in the fossil record, even as humankind in the Anthropocene is creating its own mass extinction event.

Evolutionary tinkering creates bizarre forms. As a young man, I was captivated by ancient marine fossils near the entrance of the American Museum of Natural History. These fossils are examples of some of the first multicellular (metazoan) life. They are a sparse remnant of an ancient, nearly wholly extinct, ecological assemblage alien to life as we know it.

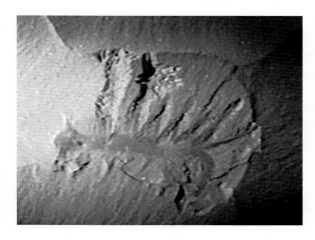

Hallucigenia fossil. *From https://upload.wikimedia.org/wikipedia/commons/1/1f/Hallucigenia_smithsonian.JPG.*

The Burgess Shale, now uplifted high in the Canadian Rockies, was once an under-sea reef. It is part of a terrene, a piece of earth's crust fused to the Western margin of the North American plate, but was originally located more than a thousand miles south. It was carried northward at a rate of centimeters per year by movement of the Juan de Fuca Plate and then accreted to North America. Subduction of the Juan de Fuca Plate along the Cascadia Subduction Zone creates high mountains, recur-rently erupting volcanoes, and great earthquakes and tsunamis in the Cascadia region of Northern California, Oregon, Washington, Southern Alaska, and British Columbia, and via the Burgess Shale this shifting of the world's crustal plates has incidentally given us a window into life as it existed 500 million years ago.

The Burgess Shale, British Columbia, Canada.

In *Wonderful Life: The Burgess Shale and the Nature of History* (1989) Stephen J. Gould recounted how Simon Conway Morris overturned a more prosaic analysis of the Burgess fossils advanced by their discoverer, Charles Walcott, "the greatest paleontologist of his day" (Gould). When Walcott, head of the Smithsonian Institution, first saw the Burgess Shale fossils, he immediately understood their extraordinary nature. *Anomalocaris* had a mouth like a circular nutcracker. *Opabinia* had five eyes. *Hallucigenia* was unworldly. But all of these Walcott classified as arthropods or worms.

In the animal kingdom, 35 (give or take a few) of Haeckel's phyla, representing distinct body plans, are generally recognized. All seem to have appeared nearly simultaneously in the Cambrian, some 500 million years ago. The question raised by Morris and his mentor Harry Whittington by their analyses of the Burgess Shale fossils was whether there were once many more phyla, as distinct in nature as Chordata or Arthropoda. Gould wrote that *Hallucigenia, Wiwaxia*, and their like were representatives of distinct phyla. This is where something went wrong with Gould's accurate insight that random catastrophe has played a major role in shaping life on earth. Morris revised his views about the phyletic distinctiveness of Burgess Shale fossils and wrote his own book, *The Crucible of Creation*. Scientific consensus, having been reversed by Morris, circled about again to Walcott. In "Showdown on the Burgess Shale" (in *Natural History*) Gould and Morris not only argued that the other was wrong, but seemed to be attempting to exorcise inner demons, Morris having reversed himself on the meaning of the fossils and Gould finding himself in the strange position of arguing with the authority on the subject, and having previously praised Morris' insight. In the end, the remarkable life-forms represented in the Burgess Shale may not represent phyla, the definition for which is blurry and more difficult to apply to long-extinct organisms. However, their fossils are a legacy of a time when life exploded in diversity and a reminder of the capricious toll of extinction, a fact that had mainly eluded Darwin, who based his theory on natural selection within and between species. The Cambrian extinction, some 488 million years ago, may have been caused by an ice age or a flood basalt eruption in what is now Australia, but this mass extinction and others were probably caused by a physical event, even if it was a physical event with biologic underpinnings.

DNA, long lost, could tell us more about the creatures represented in the Burgess Shale. Molecular and morphological evolutionary change are only imperfectly correlated. Gould cut his scientific teeth on the morphology of snail shells, observing that the evolutionary trajectory of snails, and most life-forms represented as fossils, is gradual, interrupted by spurts of morphologic change earning the name *punctuated equilibrium*. By contrast, work of molecular biologists (including myself) finds constant change at the molecular level of protein and DNA sequence, even though gross morphology may be conserved.

Because most DNA substitutions are selectively neutral, and having no effect for example on the shape of a snail's shell or anything else that would affect its survival, the molecular clock advances at a speed directly proportional to the mutation rate and regardless of population size, as Masatoshi Nei realized in a flash of insight. Molecular clocks can therefore be used to date when species last shared a common ancestor, using as anchor points known divergence times based on biogeography or the fossil record. So, for example, ratites (flightless birds) of New Zealand (kiwi), Australia (cassowary, emu), Africa (ostrich), and South America (rhea) all had a common ancestor on Wegener's supercontinent.

Over millions of years genetic change in the genome of a snail is random but steady, and on the other hand a single genetic switch can drastically change the shape or color patterning of its shell. That single change can then drive other changes. The hypothesis that a single genetic switch can elicit drastic changes in morphology, size, and color has been borne out time after time using tools of molecular genetics, and by directly tracing natural variations to single genetic switches. However, and in support of the punctuated equilibrium idea, once a genetic switch in morphology has been triggered, other genetic variants more compatible with the new form are likely to be selected, optimizing the new configuration. For example, a single mutation can turn on expression of the growth hormone gene to higher levels, converting a mouse into a larger rat-like creature. To exploit its larger size, genetic variants naturally present in mice, but uncommon, can be selected to higher frequencies in only a few generations if they facilitate more rat-like social and foraging behaviors.

Related to the power of a single mutation in a genetic switch gene to drive rapid change, a common misconception about biology is that missing transitional forms in the fossil record, or lack of information on some aspect of the machinery of the cell, are fatal to Darwinian evolution. "Aha! Scientists can't explain that," therefore let us defer to a creator. It is patently true that scientists can never be certain that an outside agency, in whatever large or small way, has not intervened in the evolution of life. Entities with capacity to manipulate nature apparently find it difficult to resist doing so. In the Anthropocene, humans are transforming the material environment and, without exaggeration, creating new life-forms both by genetic manipulations and simply by hybridizing species that would never do so in nature, thereby, humans have created ligers, beefalo, wholphins, and many other new life-forms. The cama is a hybrid between camel and llama, genera that never would have encountered one another had humans not brought them together. The rise of human civilization was drastically accelerated, or even dependent, on the deliberate and inadvertent domestication—meaning genetic modification—of dozens of plant and animal species. By deliberate burning and selective planting of trees and crops, extirpation of undesirable species, and inadvertent destruction of habitat, humans drastically reshaped the environment. However, as profound as the effects of humans are, unthinking species have had even more profound effects on Earth's biota dating at least from a billion years ago, when cyanobacteria evolved photosynthesis and oxygenated Earth's atmosphere. The main problem with the theory of intelligent design is that evidence for it—beyond what has been done by humans in the Anthropocene—is lacking, whereas evidence for natural selection and the role of large-scale extinction level events is overwhelming.

Once multicellular life appeared, there was an explosive acceleration in the pace of evolution and generation of new life-forms. Given enough time for experiments of evolution, it may then have been inevitable that some type of multicellular life would evolve that would have the interest and capacity to investigate life's origin. Mathematician Kurt Gödel proved that no system can ever provide a complete description of itself. However, that is not the challenge science faces at this stage of its history, when it is mainly learning how to frame questions. Humankind is nowhere near understanding itself or posing meaningful questions about the cosmos, most of whose dimensions we do not even perceive. However, the origin of life on Earth is a much simpler question. Progress has been rapid, although we have not been at it very long, and humans have not yet visited the numerous other worlds where the conditions on ancient Earth are closely replicated or were in the past.

The most difficult puzzles of evolution are the origins of the first replicating molecule, cells, multicellular (metazoan) creatures, and consciousness. Evolutionary biologists relate the story of life on earth in different and often complicated ways; however, there may have been four main inflection points. These inventions of nature were the self-replicating organic molecule, the cell (which compartmentalized and protected self-replicating molecules from parasitic molecules and the physical environment), photosynthesis, and multicellular (metazoan) life (which compartmentalized and protected cells from cells and the physical environment). The later innovations were dependent on or strongly altered by the earlier ones.

Each transition—molecule to living molecule, living molecule to cell, cell to metazoan, and metazoan to thinking creature capable of asking questions—is a qualitatively meaningful transition point. The stage can change how we regard the status of molecule, cell, or organ derived from a person. If we extract salt from a person's sweat, do we care about the provenance? How about a serum blood protein such as albumin? Blood? DNA? A kidney? A brain? Any living person is constantly exchanging molecules with the environment and shedding cells here and there. The argument of this book is that the unpermissioned harvest or gleaning of this matter from unconsenting people, or from their graves or biopsy specimens, is wrong, and that theft of the stuff of life is a crime of a different nature than the surreptitious withdrawal of money from a bank account. The ability to trace the very origins of life back to molecules and cells helps explain why these things are different from, say, television sets and bitcoins.

The first step is usually said to have been the hardest, but perhaps not, because it was mainly the evolution of the chemistry of a self-replicating molecule that today would not be survivable because of changes in chemistry and because cellular life would rapidly eliminate naked self-replicating molecules. Biologists study life. Although life is not on a chemical basis fundamentally different from nonlife, it is widely accepted or assumed to be so. Alfred Russel Wallace said in *The World of Life*:

> *But not withstanding these marked differences, both plants and animals are at once distinguished from all other forms of matter that constitute the earth on which they live by the crowning fact that they are ALIVE; that they grow from minute germs into highly organized structures; that the functions of their several organs are definite and highly varied, and such as no dead matter does or can perform; that they are in a state of constant internal flux, assimilating new material and throwing off that which has been used or is hurtful, so as to preserve identity of form and structure amid constant change. This characteristic rebuilding of an ever-changing highly complex structure, so as to preserve identity of type and at the same time a continuous individuality of each of many myriads of examples of that type, is a characteristic found nowhere in the inorganic world. (p. 8)*

The idea that life was a fundamentally different type of matter or was powered by a unique force not found in inorganic stuff exerted such a powerful grip on the imagination of leading scientists that in the first half of the 20th century some of the world's leading physicists embraced biology and genetics, hoping to identify that principle. As T. H. Huxley expressed it in *An Introduction to the Classification of Animals* (1869), "…life is the cause and not the consequence of organization. That this particle of jelly is capable of guiding physical forces in such a manner as to give rise to those exquisite and almost mathematically arranged structures – being itself structure-less and without permanent distinction or separation of parts – is to my mind a fact of the profoundest significance." (p 10). Huxley was on target; however, the fact of profound significance was that even that microscopic, jelly-like cell contained a DNA instruction set and so the underlying instruction set was there all along, although invisible. Huxley would certainly have used our modern knowledge of genomics and neuroscience to buttress and flesh out his materialistic view of the mind: "the thoughts to which I am giving utterance are the expression of molecular changes in that matter of life which is the source of our other vital phenomena."

Many Christians believe that other species are soulless things placed on earth to serve humankind. However, in the eons before formal religions, most people believed that animals, plants, and certain places were animated by spirits, and were bridges to sacred dominions of spirits and gods. As described by James George Frazer in *The Golden Bough*, when in the dead of winter mistletoe gleamed high above in the branches, it was because the sacred oak had been touched by lightning. The most ancient gods were ones of death and gods of renewal, like Quetzalcoatl, the feathered serpent god throughout Mesoamerica, god of the wind and with the wind, rain and renewal.

Death gods were more likely to inhabit the earth than the sky. Because practically every culture had a death god, or many, they were and are numerous, for example (as can be substantially drawn from a roster assembled by Manfred Lurker): Anubis (later replaced in Egyptian mythology by Osiris, lord of the Egyptian underworld—genetic testing now indicates that Anubis was probably not a jackal, as usually depicted, but a golden wolf), Batara Kala (Javanese god of death, light, and time), Hine-nui-te-pō (Maori ruler of the underworld, and goddess of night), Donn (Celtic), Erlik (Turkish-Mongol lord of evil and the underworld; judge of the dead), Nga (Siberian Nenets people's god of death), Ereshkigal (goddess of the Sumerian underworld married to Nergal, god of pestilence), Hades (Greek), Erebus (Greek god of primordial darkness), Ogunbali (African Igbo), Uacmitun Ahau (Mayan), Camazotz (the Mayan death bat), El Tio (worshiped to this day in Bolivia), Mictēcacihuātl and Mictlāntēcutli (Aztec lady and lord of death, respectively), Mot (Phoenician god of death and ruler of the underworld), Muut (death owl of the Cahuilla), Sedna (Inuit sea goddess who rules over Adlivun, an underworld in which the dead gathered by Anguta, her father, must pass a year), Ta'xet and Tia (Haida gods of violent and peaceful death, respectively), Tuoni (Finnish god of Tuonela, the underworld), Kalma (daughter of Tuoni and denizen of the kalmisto, Finnish word for cemetery, accompanied by the death creature Surma), Yama (Hindu and Buddhist god who rules the purgatories and judges the dead in the cycle of rebirth), Izanami (Japanese goddess of death and creation and ruler of the realm of the dead), Jabru (Elamite god of the underworld), Vichaama (Incan god of death), and Supay (ruler of the Incan underworld. See Manfred Lurker (29 April 2015). *A Dictionary of Gods and Goddesses, Devils and Demons.* Taylor & Francis. pp. 128–. ISBN 978-1-136-10628-6. Most of these gods and goddesses of death also had attendant demigods, such as the Greek psychopomps who ferry the dead.

Quetzalcoatl, Mesoamerican feathered serpent god of the wind, and with the wind, rain and vegetative renewal, sometimes depicted as eating a man.

Anubis, protector of graves, and who weighs the hearts of kings against a feather.

One purpose of this list, other than the sheer entertainment value of language and imagination, is to illustrate that the roles of death gods vary. Some are evil, horrible to look upon, and feared. Others are melancholy. Several are responsible for the cycle of life. Some were integral to the primordial creation of the world and its people.

Regardless of the nature of their death god, adherents of most religions, including Buddhism, Shintoism, Hinduism, Jainism, and religions of American Indians and those of aboriginal societies worldwide, hold that there is an innate life force and spirit in all animals, and frequently in plants as well. In Shintoism, *kami* live in animals, plants, and people, and also in natural places such as stones, mountains, and rivers…even in the dead themselves. In Japan, some 80,000 *torii* archways demarcate the world outside from the sacred world of the shrine. Through these gateways, one may access the sun goddess

Ameratsu, the goddess of the rice harvest Inari, and many others—Izanami, perhaps. In some Japanese homes, offerings to a family's ancestors are placed on a special shelf. In Hinduism, belief in the cycle of rebirth (*samsara*) and the spiritual nature of life requires one to avoid killing or injuring living things (*ahimsa*), all plants and animals being permeated by divinity.

In Buddhism, belief that other sentient creatures are reincarnating spirits also leads to kindness towards them, but the root problem that Buddhism attempts to address is (wait for it) the suffering caused by the endless cycle of rebirth and reincarnation.

Jainism, another offshoot of Hinduism traditionally known as Jaina Shasana or Jaina dharma (Sanskrit:जैन धर्म), with millions of adherents worldwide, prescribes *ahimsa*—a path of nonviolence—towards all living things, holding all forms of life equal and spiritually independent. In adherence to *ahimsa*, Jainists are strict vegetarians and may even avoid root vegetables, whose consumption could kill or injure small organisms, and avoid walking at night when they might trample small animals.

As skeptics unpersuaded by charming myth, we may scoff at the 18th-century idea expressed by leading scientists, an idea still widespread in the 21st century, that living things have a life force or spirit. However, one reason to treat all life, even the seemingly insignificant or annoying, with respect is that science continually poses more questions about life even as it provides answers. Perhaps the most important principle of Jainism is anēkāntavāda, which teaches that reality is perceived differently from various points of view and that only beings with capabilities beyond that of humans—omniscient beings known as *kevalins*—can accurately perceive reality in all its aspects. Jainism's seven propositions of knowledge, or saptibhaṅgī, are

1. *syād-asti*—in some ways, it is;
2. *syād-nāsti*—in some ways, it is not;
3. *syād-asti-nāsti*—in some ways, it is, and it is not;
4. *syād-asti-avaktavyaḥ*—in some ways, it is, and it is indescribable;
5. *syād-nāsti-avaktavyaḥ*—in some ways, it is not, and it is indescribable;
6. *syād-asti-nāsti-avaktavyaḥ*—in some ways, it is, it is not, and it is indescribable;
7. *syād-avaktavyaḥ*—in some ways, it is indescribable

Jainism's elegantly developed conception of partial knowledge can be compared to that enunciated by Donald Rumsfeld (1932–), US Secretary of Defense (1975–1977): "There are known knowns. These are things we know that we know. There are known unknowns. That is to say, there are things that we know we don't know. But there are also unknown unknowns. There are things we don't know we don't know." The identity and source of the first self-replicating molecule remains mysterious.

From molecule to life

Darwin had much to say about how life evolved, but little about its emergence. In a letter, Darwin speculated that life began in "a warm little pond," but it is fortunate that he did not concern himself with the chemistry because in Darwin's time the molecules, the extreme antiquity of life, and the chemistry that probably existed on early earth, were all unknown. Darwin's evolution must have started with Dawkins' replicator molecule, but which one—DNA, RNA, protein, or some other?

A water molecule forms elaborate crystalline structures, and is essential to life, but does not evolve. Self-replicating molecules are abundant in nature but to form the basis of life the molecules also had to be built of readily available elements, somewhat resilient to degradation, and capable of further evolution. Also, it is the nature of self-replicating nucleic acids that they require compartmentalization to protect themselves from predatory and parasitic molecules, as could have happened in several ways, as simply as lipid bilayers form in an emulsion of nonpolar lipid and polar water.

What are the elements that are the building blocks of organic molecules? Ninety-eight elements occur on Earth, but only a handful—hydrogen, oxygen, nitrogen, phosphorus, and sulfur—are abundant in our bodies, and several other trace elements such as iron and copper are essential. Minerals are pure or nearly pure forms of elements. In nature nonorganic carbon occurs nearly pure in different forms, including graphite, anthracite coal, diamond, and buckeyballs. Atomic elements are organized in the periodic table by the numbers of protons in their nuclei and by the electrons in orbitals around the nucleus, enabling their chemical bonding with other atoms, to form molecules. Atoms are themselves products of physical processes of assembly and atomic decay. All carbon has six protons (the atomic number), but carbon can have two extra neutrons beyond the normal complement of six (C-14 carbon). The extra neutrons do not alter the chemical behavior of carbon but occasionally one of the extra neutrons suddenly decays into a proton and releases energy as a high-energy electron. In this process, known as beta decay, a nitrogen atom is made from carbon, and in this way the ancients were correct that elements are transmutable.

Crystals are a clear example of molecular self-replication and self-assembly. Water molecules self-assemble into ice crystals. Carbon atoms self-assemble into macroscopic, ordered macromolecules including graphite and diamonds. A common misunderstanding is that large-scale order cannot arise out of chaos because of the difficulties of this happening by random chance, or due to the second law of thermodynamics. We have already seen how Wallace's and Darwin's natural selection is the driving force of evolution, and that organic life-forms are neither the product of random events or an engineer's blueprint. However, complex, ordered nonorganic structures are also constantly forming, at the expense of increasing entropy around them. More so than DNA code, all life depends on the capacity for molecular self-assembly, and the structure of the planet and the universe are entirely due to it. Haeckel was wrong that the cell walls of cork formed by crystallization; however, practically every complex structure of our cells owes its existence to the ability of organic molecules to fold and otherwise arrange themselves in ordered structures, and as will soon be seen, such molecules can—under the right conditions—take on a life of their own.

Anyone who has ever mixed nonpolar oil with polar water knows that the two spontaneously separate, first in cell-like droplets and, if left long enough, into oil and aqueous phases. A soap bubble membrane blown by any child is similar to a cell's membrane in its lipid bilayer structure, although the cell membrane is also embedded with functional components such as receptors and pore complexes. The membrane of a cell or soap bubble is largely composed of simpler molecules—phospholipids that are amphipathic, having a hydrophobic, nonpolar end of the molecule that spontaneously orients towards the interior to interact with other nonpolar chemical groups, and a polar end that orients outwards, to interact with water and other polar molecules. The cell membrane and many other semispontaneous molecular assemblies within a cell can be so large as to be visible to the naked eye. For example, the crystalline lens of the eye is made of millions of crystallized protein molecules, often evolutionarily repurposed from other functions but then specially evolved, the better to make a durable lens with good optical qualities.

Self-assembly can also involve covalent bonding between atoms and molecules, as will be familiar to anyone who knows of polymerization. In polymerization, very large durable structures such as sheets and fibers can spontaneously assemble from reactive ingredients, and the reaction can be exothermic or endothermic, depending on the polymer. On a simpler level, the chemistry of covalent bonding causes water molecules to self-assemble from two elements, causing the hydrogen (H_2)-filled Hindenburg to explode. Water is a hetero-elemental molecule consisting of three atoms (H_2O). A tear in the zeppelin's fabric released large quantities of H_2 and one spark was sufficient to trigger its massive exothermic reaction with oxygen (O_2) in the atmosphere. Why? Because elemental hydrogen and oxygen can share electrons in free orbitals around their nuclei, leading to a lower free energy state than existed as O_2 and H_2O. How did the oxygen and hydrogen molecules find themselves? By random collision enhanced by heat. Once started, the exothermic reaction did not stop until most of the hydrogen was consumed or dispersed by the force of the explosion. Before petering out, trillions of new water molecules had "self-assembled" and by now dispersed, becoming part of earth's oceans, atmosphere, and people.

The explosion of a hydrogen-filled zeppelin increases disorder (entropy) in an obvious way, disrupting travel plans as well as releasing massive amounts of heat. The obedience of crystallization to the laws of thermodynamics is less obvious but no less real. When atoms or molecules crystallize because of the atomic forces between them, the gain in order is more than offset by the release of heat, which represents the thermal randomization of the surroundings. Crystals, polymers, and life-forms are local islands of decreased entropy created at the expense of increased entropy around them.

The question is, which of three molecules—DNA, RNA or protein—was the first self-replicator going about creating order at the expense of disorder all around it?

DNA: DNA is a very large heteropolymer, the human genome being composed of some 3 billion nucleotide "monomers" assembled into 23 large polymers, i.e., the chromosomes. As has been discussed, DNA consists (primarily) of four DNA bases, with many secondary (epigenetic) modifications. DNA normally occurs as a heteroduplex, the two strands constructed of covalently bonded DNA bases being bound to one another noncovalently, by hydrogen bonding between complementary base pairs A:T and G:C.

In the laboratory, DNA can be synthesized such that the sequence of nucleotides is random. However, in life, one strand serves as a template for the synthesis of a daughter DNA strand that, barring mutation, faithfully replicates the sequence of the parental cell. Cells use a complex cassette of cofactors and enzymes to inhibit DNA synthesis when it is not wanted, to nudge it forward when it is required, and to audit and repair the inevitable errors that occur either before, or as most often happens, during, replication of a DNA polymer with 3 billion elements.

The heteroduplex nature of DNA is key to its functions, because if the two strands unzip, either can serve as the template to generate a new DNA strand, as James Watson and Francis Crick pointed out in the last sentence of their paper describing the discovery of the structure of DNA. The heteroduplex nature of DNA is also essential for its ability to be transcribed into RNA. The DNA molecule is a replicator molecule. At some point very early on, DNA became the master. Except for some RNA viruses that use RNA as their primary genetic material, DNA carries the genetic instructions for life on earth. However, unless life first evolved on a planet other than Earth, it is more likely that the first replicator molecule (again borrowing Richard Dawkins's term) was not DNA. This molecular precursor of all life on Earth would have mainly been capable of making high-fidelity copies of itself, with occasional mistakes (mutations) to serve as genetic grist for natural selection.

Transposons—self-replicators within the genome: The genomic revolution which began in 2000 with sequencing of the human genome and the genomes of other species has led to a deep understanding that much of the human genome is an evolutionary artifact. In addition to 25,000 or so functional genes are thousands of pseudogenes, which are gene-like sequences that have undergone some sort of mutation so that they have lost function. These pseudogenes can be viewed as fossil genes still distinguishable in the genomic stratum but whose identities are slowly being eroded by the accumulation of mutations, there being no selective force to maintain their sequence fidelity. In a functional gene even a single base substitution—one letter in the DNA code—can have catastrophic results. The existence of a large complement of independently replicating and usually destructive transposable elements in our genomes is also counterfactual to a scheme in which human evolution was ordered and guided. These transposons, which are themselves selfish independent genetic elements uninterested in human welfare, are relatively small pieces of DNA that possess an enzyme, transposase, that enables them to excise themselves from one location and reinsert elsewhere. In maize, where Barbara McClintock at Cold Spring Harbor discovered them, the insertions of transposable elements lead to color variations between kernels.

RNA: At its origins, life may have been an RNA world (Walter Gilbert, *Nature*, 319: 618, 1986, Joyce and Orgel, 2006), with RNA executing enzymatic and structural functions, replicating on its own and then later using protein as its servants and itself becoming servant to DNA.

Like DNA, RNA is also a nucleic acid heteropolymer constructed of four nucleotide bases and many epigenetic modifications of these bases. Unlike DNA, RNA is usually single-stranded. However, RNA usually folds back on itself to form highly complex structures. The complexity and multiplicity of RNA structures enables RNA to take on some of the enzymatic and structural roles that were thought reserved to proteins.

As first shown by Nobel laureate Thomas Cech in 1982, RNA can, like a protein enzyme, catalyze chemical reactions. Potentially, an RNA could have catalyzed replication of itself, using itself as a template. The first RNA enzyme discovered by Cech was one that could splice out a piece of itself—a potentially reversible reaction that would allow RNA to construct itself piece by piece.

RNA is also the connecting link between nucleic acid and protein. RNA, unlike DNA, can be directly translated into protein using the assistance of other RNA molecules called tRNAs (transfer RNAs). These tRNAs are ancient, their sequences being highly conserved, such that the tRNA sequences of people are virtually indistinguishable from those of a starfish.

Another ancient RNA is Ribonuclease P (Guerrier-Takada et al., 1983), an RNA enzyme. RNAse P is found at the center of the ribosome where it is directly responsible for the step-by-step assembly of proteins. Guerrier-Takada, Altman, and colleagues postulated that an RNase P-like enzyme might have first acted on other nucleic acids rather than proteins: "If proteins were relative latecomers in the evolution of macromolecules, then primeval manipulations of nucleic acids may have been carried out entirely or predominantly by catalytic nucleic acids themselves. The remnants of these early, important, biological events may be apparent in the reaction of RNAase P."

Decades later, speculation that RNA can self-replicate seems to have become reality. Biochemists set up conditions for artificial evolution of RNA in the test tube and by 2001 had an RNA enzyme 189 nucleotides long that could add 14 nucleotides to itself, using another RNA to specify the sequence. Within a decade, and "cheating" a bit by using a combination of in-test-tube molecular evolution and direct assembly of some of the most effective pieces of what appeared "on its own," an RNA enzyme had been created that could add 95 nucleotides of a specific sequence and actually synthesize a complete, active ribozyme (Wochner et al., 2011).

The spectrum of reactions that RNA can catalyze has been broadened by the ability to use cofactors: for example, an engineered RNA, like a natural alcohol dehydrogenase enzyme, is able to use the cofactor NAD (nicotinamide adenine dinucleotide) to metabolize alcohol. Once the RNA bases formed, there are a variety of ways, including in clays, as shown by Leslie Orgel in 1996, that they could have spontaneously condensed into long RNA molecules. In 2009 a watershed divide was passed when the first self-sustained self-replication of an RNA was achieved in the test tube, with a doubling time of about 1 h (Lincoln and Joyce, Science, 2009).

However, could the RNA ribonucleotides themselves—the ingredients to make RNA—have formed abiotically? As Sutherland said in 2009, "At some stage in the origin of life, an informational polymer must have arisen by purely chemical means." For RNA, the answer appears to be "possibly." Under conditions that probably occurred on ancient Earth, and without synthesizing the individual RNA bases first, Sutherland was able to make RNAs composed of two of the RNA bases, the pyrimidine nucleotides A and T (adenosine and thymine).

Protein: Could proteins, which are so versatile and pervasive, have been the first self-replicators? Amino acid building blocks of proteins form readily in the conditions that probably prevailed 4 billion years ago, and covalently bond with each other to form peptides. Possibly, the first self-replicating molecule—if it was a protein—did not use the same 20 amino acids. The Murchison chondrite meteorite contains some 75 amino acids, all of which presumably formed abiotically. If life evolved elsewhere beginning with a self-replicating protein, it might well have used a different palette of amino acids.

Alexander Oparin, a visionary, speculated that the atmosphere of early earth was reducing, allowing the formation of large organic molecules from the prebiotic soup of Darwin's "warm little pond." Inspired by Oparin, Harold Urey and Stanley Miller jolted the scientific community by exposing a primordial soup of methane, ammonia, hydrogen, and water to an electric spark. Nearly half of the carbon in the methane was transformed into more complex organic compounds, thereby showing that it was easy to make organic matter from inorganic under conditions that existed soon after Earth was formed. As far as they knew, only three amino acids were abiotically synthesized by Miller and Urey, including glycine, α alanine, β-alanine, and possibly two others, aspartic acid and α-aminobutyric acid, which were represented by faint spots on their paper chromatography. However, decades later, Jeffrey Bada, a former graduate student of Miller's, reanalyzed the untouched, still-sterile, contents of the original flasks. Using more sensitive gas chromatography technology developed over the intervening years, Bada discovered that 22 amino acids were present in Miller's flasks.

Not all of the amino acids synthesized abiotically are found in proteins, but decades after the Miller-Urey experiment, other scientists abiotically synthesized all of the 20 amino acids found in proteins and essential to life. They used a volcanic gas spark discharge model or a deep ocean hydrothermal vent model in which a spark was passed through a hydrogen sulfide-rich mixture. These last experiments mimicking volcanoes and deep-sea hydrothermal vents underline that harsh environments, and not just Darwin's "warm little ponds," could have been crucial for the origin of life. Early Earth had a plentitude of hydrothermal vents and volcanoes, as well as warm little ponds. One amino acid, glycine, can even form spontaneously in the vast and cold reaches of space, where it has been detected. To this day, extremophile bacteria thrive in alien environments where most life cannot, in hot springs, hydrothermal vents, and in rock; for all we know, life got its start in some environment that was very hot and toxic, or forbiddingly cold.

If a protein catalyzing synthesis of itself was born abiotically in some hard or harsh place or survived passage to Earth in a meteor, it could diffuse or be carried here and there throughout the world. Wherever it would find suitable chemistry, it would tend to make itself more abundant, leading to its further dissemination. This possibility has been made more plausible by the discovery in 1982 of tough, self-replicating proteins predicted to be the cause of transmissible spongiform encephalopathies some two decades previously by Alper and Griffith, and dubbed prions by their discoverer, Nobel laureate Stanley Prusiner. The prion precursor protein has 209 amino acids and is expressed in neuronal membranes.

Very much like the crystalline nidus that causes sudden freezing of a supercooled liquid, a prion is infectious, triggering misfolding of normal prion precursor proteins that otherwise do not misfold. The misfolded protein forms highly structured amyloid fibers that self-replicate within neurons by catalyzing the misfolding of more and more precursor protein. Furthermore, in its new conformation, the prion protein, which can now be called a prion, is tough and survivable.

Fortunately, and due to sequence specificity required for misfolding, and minor differences in sequence across species, prion diseases are mainly transmitted within species. Lev Goldfarb, at NIH, discovered that some people with Creutzfeldt-Jakob disease carry genetic variants of the prion precursor protein that can cause the protein to fold into a prion. Within-species transmission can follow transplantation, injection of growth hormone derived from an infected cadaver, or consumption of tissue from an infected patient. Transmission is likely following such exposure, and deadly, but the disease can be delayed, occurring a mean of 12 years but as long as 50 years after exposure, and as influenced by a person's genotype. At the NIH, Nobel laureate Carleton Gajdusek discovered that Micronesians suffered the neurodegenerative disease kuru if they ate the diseased brains of dead relatives as a ceremony of respect and mourning. Among the Fore people, the disease was epidemic, being the most common cause of death among women of affected villages (Collinge, Alpers et al., Lancet, 2006).

Unfortunately, prion diseases can be transmitted cross-species. Variant Creutzfeldt-Jakob disease—an uncommon cause of dementia—is a prion disease that can be transmitted not only within but between species. Tens of thousands of cows were affected by mad cow disease in Britain, at least 184,000 by 2011, plus cows in other countries. A handful of infected cows were found in the United States but the true number is unknown. When mad cow disease was discovered, the UK government insisted that British beef was safe, but this was not true. Prion was found to have caused variant Creutzfeldt-Jakob disease (VCJD) in some 160 people in the UK, plus other cases in countries worldwide (Collinge, Alpers et al., Lancet, 2006). Millions of cattle that could not be consumed as food had to be destroyed. It is questionable whether milk from infected cows is safe, as authorities have claimed, because prions have been found in milk from infected animals. Recently, as may give hope for people at risk for prion diseases, the onset of transmissible spongiform encephalopathy has ingeniously been blocked by Soto, Diaz-Espinoza, and colleagues with an antiprion protein that itself self-replicates (Molecular Psychiatry, 2018).

After all this about prions, it is certain that the first self-replicator molecule was not a prion of the exact type causing neurological disease in humans, because these prions act in a cellular environment and by converting normal proteins of considerable length. However, prions demonstrate that a protein can be tough, independently survivable, and infectious. As shown by Prusiner, prions in manure from infected cows are infectious, and prions can remain dormant in the soil for extended periods. The survivability and ability to self-replicate explain why prions are infectious, mainly causing neurodegenerative disease. However, other types of prions affect other cells, organs, and species (including fungi and mink). For example, in humans, α synuclein can misfold to cause multiple system atrophy (MSA). Dozens or even hundreds of human proteins have prion-like domains that make them capable of autocatalytic misfolding. Interestingly, several proteins that bind to and work in close partnership with RNA have prion-like domains. Proteins with prion-like domains are found throughout nature, and study of their properties, capacities, and interactions with RNA could take science toward an explanation of how the first self-replicator could have been a protein.

Although the identity of the first self-replicating molecule remains unknown, the idea of self-replication of nonlife is not fantastical. Superficially, computer and biological viruses are examples of simple self-replicating machines, but each hijacks a much more complex machinery to achieve its purpose. For example, a computer program that outputs its own code, known as a quine, can be very simple: a = "a = %r;print a%%a";print a%a. The DNA equivalent of a quine is the transposon, a small piece of DNA surrounded by palindromic repeats that can make copies of itself, inserting into new places in the genome. Viruses are more complex than transposons. A virus is to a transposon as a computer virus is to a quine. The viral sequence may integrate into the host genome, but also has the machinery to export itself in packaged form outside of the cell and to infect new cells. Just as software engineers design protections against computer viruses and quines, so does the cell attempt to protect itself against transposons and viruses. However, each of these self-replicators can survive only in the context of a more complex system, living or computer. Importantly, and as compared to other molecular "half-life" discussed in this book, such as prions and viruses, transposons are not as readily transmitted horizontally, from individual to individual or from one species to another. However, while horizontal transmission is rarer for transposons, it has also apparently been an important feature of transposon evolution, with examples of transposons invading the genomes of new species, and then spreading by vertical transmission from generation to generation, and then spreading within the genome of the individual. One such example is the P element, which Margaret Kidwell discovered to have invaded the genome of the fruitfly *Drosophila melanogaster*, but also another fruitfly species. Transposon invasions have strongly shaped the genomes of humans and other vertebrates, not just fruitflies. Recently, a survey of more than 300 vertebrate genomes by Hua-Hao Zhang and others revealed nearly 1000 independent instances of horizontal transmission of transposons across species.

Evidence of life on Earth appears in 4.1 billion-year-old Australian rocks, and definite fossil evidence of life on Earth, in the form of bacterial mats (stromatolites), is found in rocks 3.5 billion years old. This is "only "a billion years after Earth's formation, and during much of that time the earth was probably suffering bombardments of large meteors that might have been capable of killing most or all life. In the Archean, from 4 to 2.5 billion years ago, life formed quickly but evolutionary change thereafter from single-cell life to multicellular life was slow, and dependent on chemistry. A main limitation was lack of oxygen, the molecule that modern aerobic organisms mainly reduce in order to oxidize, and utilize, organic compounds. Instead, like some organisms that exist in anaerobic environments today, Archean organisms used sulfate (SO4). Sulfate was also in short supply, but a reliance on sulfate by early life has been demonstrated by the selective biotic depletion of isotopes of sulfur (^{34}S and ^{33}S) in rocks formed during the Archean period (see, for example, Zhelezinskaia et al., 2014). The first photosynthetic organisms evolved and cyanobacteria began oxygenating Earth's atmosphere and oceans at the end of the Archean eon, about 2.4 billion years ago. However, even then, and as determined by the ratio of chromium (Cr) isotopes in ironstone, a sedimentary rock, the heavier ^{53}Cr isotope being more likely to be abiotic oxidized and weathered than the ^{52}Cr isotope, it was not until about 800 million years ago that oxygen levels *may* have risen to anything like present-day levels (Lyons et al., 2014). When oxygen levels finally did rise, about 800 million years ago, the Cambrian explosion of metazoan (multicellular) life followed almost immediately.

There are many things we do not know about the evolution of life, but the history of life provides compelling illustrations of the ability of adaptive selection to make new life-forms, beginning with the stuff of chance and circumstance. For example, flight evolved many times, in plant seeds, insects, spiders, and even fish. Among terrestrial vertebrates, lemurs, squirrels, lizards, frogs, and dinosaurs all independently evolved flying/gliding forms, namely the honey glider (*Petaurus breviceps*), the flying squirrel (genus *Glaucomys*), the flying dragon (genus *Draco*), and Wallace's flying frog (*Rhacophorus nigropalmatus*). Their evolutionary derivations are evident at many levels: anatomic, molecular, physiologic, genetic, and biogeographic. Less obviously, birds evolved from dinosaurs and bats shared a common ancestor with carnivores and artiodactylids (hoofed animals). Squirrels and lemurs and lizards climb and leap from branch to branch. Over time, some became specialists in leaping and progressive extension of the skin between their limbs and body gave them an advantage. A flying squirrel might be on its way to becoming a bat, except that bats occupy that evolutionary niche. Flying squirrels and honey gliders were able to evolve towards the same evolutionary niche because one species' ancestors were in North America and the other's were in Madagascar.

The array of social interactions between individuals of the same species and between individuals of different species was only achievable because of the capacities of metazoan organization. Metazoan life exploited thousands of ecological niches unavailable to the single cell. Each cell, rather than being an independent generalist that is a little bit good at everything (think of a promising recruit with a good general education), can differentiate, beginning from a relatively small squadron of pluripotent cells into an army of specialists. Working together, metazoan cells could build complex organs enabling communication, locomotion, immune defense, emotion, the efficient harvesting of food resources, and cognition. Delicate tissues were housed in tough integuments and supported by exoskeleton, cartilage, or bone.

Individually, cells are stupid. If stupid cells can work together to build something so complex and elegant, why cannot people? They are genetically programmed using a relatively simple instruction set. Single-cell organisms are remarkably self-sufficient. However, the cells of metazoan organisms are interdependent. Individual cells often fail miserably in their functions and may be rapidly discarded through programmed cell death. Often the best thing a cell can do in the service of a multicellular organism is to die. Together, cells do unimaginably complex things, while working on its own one cell can do practically nothing. Taken out of the body, cells can be kept alive only if cultured under special conditions. Only the lowly sperm survives for a time outside the body, and not for long—it needs to find its way into the special environment created by a receptive female, or its lifespan, short as it is under the best of circumstances, is prematurely shortened.

Epigenetic origins of cellular, and interindividual, diversity

A key to the origin of metazoan, multicellular life was the evolution and refinement of systems for cellular differentiation and specialization. The epigenetic mechanisms that enable this cellular specialization also have a very important role in the individuation of people in response to environmental exposures.

In most people, and except for some people who are chimeras and recipients of transplants, all cells in the body derive from one primordial germ cell. Subsequently, the cells derived from this germ cell by successive divisions have virtually the same genotype, except for sporadic mutations, as can play a very important role in tumorigenesis. Genetic identity of cells throughout the body explains why a forensic DNA profile (fingerprint) from a semen sample can match that derived from blood or saliva. However, the body needs more than one cell type.

Pluripotent stem cells differentiate into thousands of cell types the body needs. How? (see Goldman, Schuebel, Gitik and Domschke, "Making Sense of Epigenetics," *International Journal of Neuropsychopharmacology*). More than half of our 25,000 genes are expressed in almost all cell types. However, differentiation of germ cells into the major germ layers—endoderm, mesoderm, epiderm—and the myriad specific cell types represented in these three tissue domains is primarily accomplished by epigenetic change. The epigenetic changes primarily involve secondary modification of DNA nucleotides, and modifications of chromatin proteins that package DNA.

The key difference between epigenetic change and genetic change is reversibility. A DNA (genetic) mutation is difficult (but not impossible) to undo, but epigenetic change, whether in DNA or the protein packaging of DNA (by histone proteins) can be readily undone—explaining why it is relatively easy to make an induced pluripotent stem cell (iPSC) from a lymphocyte in peripheral blood or a fibroblast in skin, but more difficult to transform a rhinoceros into a rhinoceros beetle or an elephant into an elephant shrew.

The epigenetic modification of histones in a local DNA region can open that region for access to transcription factors, poising genes within it for transcription to RNA. Many of the transcription factors that bind to sequence motifs in the DNA region are themselves cell specific, and alone or in combination they can enhance or suppress transcription, or lead to the expression of alternative transcripts from different transcription start sites. All of this complexity is partly why the gene concept is blurred. RNA molecules have a variety of functions, including structural, regulatory, and enzymatic, and are templates for synthesis of proteins at the ribosome. The web-work of molecular interactions that make different cell types distinct, and that represent further levels of epigenetic regulation, is complex: for example, there are at least 10 times as many proteins encoded by genes as there are genes.

Before and after birth, humans and other species can be epigenetically reprogrammed. Genetically identical human twins are far more likely to be similar than siblings sharing 50% of their genes by descent. However, epigenetic events can lead to dramatic behavioral differences. In genetically identical rats, Michael Meaney (1985) showed that whether or not neonates are licked by their mother dramatically altered their subsequent emotional development, as was mediated by epigenetic change at the glucocorticoid receptor gene. In primates, early maternal care is essential for normal emotional development, and as was proven by maternal deprivation experiments on rhesus macaque monkeys by Harry Harlow, and subsequently, Steven Suomi. These findings, and controversy about maternal deprivation experiments in non-human primates, have been discussed elsewhere (Goldman, 2012).

Via epigenetics, the social role of individuals of many species, for instance ants, bees, and mole rats, is directly determined by environmental exposures: for example, the feeding of so-called royal jelly, which programs the expression of different constellations of genes. In primates and other species, social position itself has consequences both dramatic and subtle for the ways genes are expressed, further influencing social status. The effects of social stress on gene expression in people have been well-documented. These studies might be confounded in a variety of ways, because it is not ethical to deliberately change a person's social status to study epigenetic change, and people inhabit many types of social hierarchies. Socially subordinate monkeys are harassed the most, groomed the least, and must constantly be on the lookout, such as when they have found tasty food. Merely changing the social position of female rhesus monkeys to make them subordinate upregulates immune-related genes that promote inflammation, biasing immune function towards chronic inflammation, and—potentially—premature aging (Snyder-Mackler et al., 2016). Women from stressful, impoverished environments can respond by chronically activating the hypothalamic–pituitary stress axis, releasing cortisol, as is well known, but also secreting increased amounts of neuropeptide Y. The *NPY* gene has a common functional polymorphism in its promoter region that alters anxiety responses (Zhou et al., 2008). However, the effects of elevated neuropeptide Y secretion also extend beyond behavior, to metabolism. As the late Zofia Zukowska (Kuo et al., 2007) discovered, stress-induced secretion of NPY leads to obesity and metabolic syndrome, and Finnish scientists (Karvonen et al., 1998) found an amino acid substitution (missense variant) polymorphism of NPY that alters the function of this peptide and increases serum cholesterol and lipoprotein levels associated with obesity.

Via evolution, and with deep roots extending back to the first multicellular life, human capacity has evolved, including capacity for adaptation to a range of environments and roles. Neurogenetic variability, maintained by natural selection, enhances adaptability of human populations across time and space.

Summarizing, nothing that has been discovered about life so far is incompatible with it having self-originated, self-assembled, via the action of chemistry, mutation, natural selection, and chance. Whereas our evolution was probably not directed, we have evolved the capacity to self-direct it, and on a smaller scale, to self-direct our own lives. This capacity is profoundly influenced by the social milieu and the social contract. In combination, genetic variants, interacting with environmental experience that finds its way into the genome via epigenetic change, make people profoundly unequal. People may make wise choices later in life, but they do not choose their parents, who supply their genetic inheritance, nor do they choose their early life experiences or later environmental exposures that may profoundly alter physiology and behavior. In life, we play the hand dealt to us by evolution and happenstance of birth, but some are dealt hands that are a challenge to lose, while others are holding hands that can be won only with skill and patience, or not at all. In complexity and capacities, metazoan life, and especially humanity, is far more than the sum of its parts. Yet, our parts are special. Even as the boundaries between life and nonlife, between human and nonhuman, and between self and other have been blurred, it can be recognized that derived pieces of persons have a special meaning in their function during life, the differences between us and in how life originally evolved.

10 ⊚

Viruses and other half-life

Anyone who has watched a zombie movie has probably asked themselves whether it is possible to kill something that is not truly alive. At the boundary of life and unlife are entities such as viruses that may be best called half-life. This chapter examines how viruses inhabit that boundary, and along a sliding scale that includes entities that are self-replicating but even simpler and less "lifelike," such as transposons and prions, versus intracellular parasites and bacteria that, while far more lifelike, are also easily understood as simple automata.

First, it is intellectually honest to pay respect to the transformative effect of viruses and other germs on our multicellular life and drop the anthropocentric view that we are evolutionarily superior. Humans are objectively more complex, but germs are equally perfected by evolution and objectively more ancient. The COVID-19 virus, SARS-CoV-2, probably was introduced to the human population via one transmission event from a reservoir population in bats, perhaps via the pangolin, which is a cat-sized, scaly anteater that hunts ants and termites in the trees and that attempts to defend itself by curling up into a defensive ball. Unfortunately for the pangolin, its meat is considered a delicacy in some East Asian countries and its scales are prized as a medicine in China. Against humans, the pangolin's ball strategy is ineffective. Every year, and this will probably go on until pangolins are exterminated, hunters kill about 100,000 pangolins, often capturing the balled-up pangolin simply by picking it up and throwing it into a bag. Hundreds of tons of pangolin scales, which are mostly keratin (like hair), are shipped to China every year, and the scales, pangolin, and pangolin meat arrive in places such as the Wuhan market that was near the epicenter of the COVID-19 pandemic. Or, perhaps the virus was inadvertently released from a laboratory in Wuhan. If SARS-CoV-2 did make the jump from pangolins to humans, it is ironic that pangolins have suffered so much from people, largely for the inane reason that people think that its scales are medicinal and that at high magnification the SARS-CoV-2 virus looks exactly like a ball.

Whatever its true origins, COVID-19 has spread to every continent save Antarctica, infected tens of millions, killed hundreds of thousands, crashed economies of all nations, and changed the way people live. The main defect of the narrative of COVID-19 is that throughout the pandemic people, including purported experts, wrongly claimed that COVID-19 and its consequences were unprecedented in severity and social impact. Probably, this tendency to forget anything distasteful that happened to previous generations is strongly related to our lack of preparedness for COVID-19. In the past, hurricanes, earthquakes, tsunamis, meteors, and germs have killed millions, but the newest hurricane or germ is always deemed the worst ever. As my mother liked to say, "I've seen worse," and if she was channeling Gaia, she had.

Immortal. https://doi.org/10.1016/B978-0-323-85692-8.00010-1

Each pandemic is unique in nature, and we will make distinctions between pandemics caused by viruses versus other germs, but as Hans Zinsser wrote in *Rats, Lice and History*, pandemics have, over and over again, shaken and destroyed civilizations. Germs brought down empires in Europe during medieval times. In the Americas, germs enfeebled and destroyed indigenous empires and were foundational to European colonial empires during the great age of European exploration and exploitation. Jared Diamond in *Guns, Germs, and Steel*, and historian Alfred Crosby who coined the phrase "virgin soil epidemic," illuminated how, time after time, diseases endemic and epidemic in vast Eurasian populations geographically interconnected at common latitudes destroyed geographically and demographically isolated indigenous peoples who were immunologically naïve. In modern times, a pandemic can spread to any distant part of the planet in days, but in ancient times pandemics traversed interconnected populations over the course of years, often to circle back again and again.

Pandemics change the nature of governments, providing an opportunity for authoritarians from all parts of the political spectrum. Frank Snowden, in *Epidemics and Society: From the Black Death to the Present*, proposed that the Black Plague contributed to centralization of power by governments, with governments imposing quarantine. As anyone who endured COVID-19 shelter-in-place orders knows, pandemics, like wars, enhance the power of the state at the expense of individual freedoms. Pandemics lead some people, who may be sympathetic to a repressive regime, for example China, to praise that state's authoritarian model. Probably China had fewer COVID-19 deaths on a per capita basis than most democracies, but on the other hand, the repression of that nation was in the first place the main reason why the pandemic was not stopped at the outset, which is the moment to stop pandemics. Pandemics change our reactions to others, permanently increasing xenophobia and, for example, leading some people to curse anyone who looks Chinese, as if any ordinary citizen of China or person who appeared vaguely to be of Asian descent had anything to do with the pandemic.

Within families, pandemics increase the intensity of household interactions but further distance family members who are separated. Formerly, families were physically together, but even so pandemics could cause emotional distancing. Joshua Loomis in *Epidemics: The Impact of Germs and Their Power Over Humanity* wrote that smallpox was so devastating, with virtually everyone exposed and a mortality rate of 30% and more than 90% in young children in many countries, that parents would wait to see if their children survived smallpox before naming them. The power of pandemic, especially before modern understanding, pervaded the imagination, evoking dread, hyperreligiosity, and despair of a better future in the material world.

Although SARS-CoV-2 is known as COVID-19 (not the Wuhan virus), in recent times germs were often named for geographic origin. These labels, such as Marburg, Lassa, Rift Valley, Ebola, and Hanta viruses, or Spanish flu, were often only semiaccurate, but they were colorful and memorable. During the COVID-19 pandemic, geographic naming appears to have resumed, and for example, with more virulent variants of the virus being commonly referred to by places they were first identified, and for example, the UK and South African variants. Before the age of geographical naming, pandemic diseases were often remembered by a color associated with them. Tuberculosis was the White Plague. Smallpox was the Red Plague, or "speckled monster." Edgar Allan Poe's "Red Death," although perhaps based on a cholera epidemic he witnessed in Baltimore, was fictional, and if it was based on cholera, Poe got the color wrong.

Cholera, called the "blue death" because of how it dehydrates its victims, turning skin slate blue, was discovered only in 1883 by Robert Koch, during the fifth cholera pandemic. The bacterium makes cholera toxin, which triggers massive fluid loss by intestinal epithelial cells, leading to spread of the organism via contamination of water. To save a cholera victim's life, a liter of water balanced in salt composition must be replaced every hour, as is now done throughout the world with prepackaged salts. Koch named his bacterium *Vibrio cholera*. As I write this book, Phase 3 vaccine trials for COVID-19 are well underway, and I would hazard a guess that we will soon have an effective vaccine (and I received my second dose of Pfizer's RNA-based vaccine today, as I was proofing this book), along with the testing, contact tracing, social distancing, handwashing, and masking that we know can stop this virus in its tracks. However, does that mean this germ is going away? Probably not. We have known for more than 100 years that cholera can be prevented by providing clean water. Yet, cholera is endemic throughout much of the world, with up to 4 million cases a year, and the seventh cholera pandemic is currently ongoing in Haiti. Having been brought to that country by United Nations peace-keeping troops, cholera has so far sickened almost a million Haitians and about 1 in 100 who contracted the disease have died. The story of cholera is paralleled by other germs. Germs that are epidemic tend to become endemic – always a part of the human present and not just our past. Already, the SARS-CoV-2 virus has mutated into more virulent, and probably more deadly forms such as the UK and South African variants, and the evolutionary race between human and germs continues.

The "Blue Death" is bad, but the Black Death is worse. The Black Death arrived in Europe in 1347, brought from Asia by ships docking in the ancient port of Messina. Like cholera, the Black Death is also a bacterial disease, caused by *Yersinia pestis*. Otherwise it is totally unlike cholera. Proving that there are diverse ways for a disease to be horrifying, there are three main clinical presentations of *Yersinia pestis* infection in humans. Bubonic plague chiefly infects the lymphatic system and is the most common of the three. Swollen lymph nodes known as buboes appear in armpits, groin, or neck, giving the name to bubonic plague. Other signs and symptoms of bubonic plague include high fever, extreme weakness, abdominal pain, nausea and vomiting, bleeding, and blackening and gangrene of fingers, toes, and nose. Septicemic plague, also transmitted from fleas, infects the blood, leading to septic shock but not all the other manifestations of bubonic plague. Pneumonic plague chiefly involves the lungs. It is usually contracted by inhalation, and is infectious via respiratory secretions and not only through the secondary host of the flea. Within 5 years of the arrival of *Yersinia pestis*, an estimated 25–30 million Europeans had died, about half the population of Europe, and worldwide the population was reduced by at least a full 100 million from 475 million, taking 200 years to recover. This was not the first time the Black Plague had swept through Eurasian populations—it devastated Europe in the Neolithic era, as shown by ancient remains in Sweden from 3000 BC, and was the cause of the plague of Justinian, which first struck Pelusium, Egypt in 541 AD and reached Constantinople the following year, Rome in 543, and Britain a year after that. The Roman Empire, revitalized under Justinian, never really recovered from the Plague of Justinian and its recurrences at decade and a half intervals, in 558, 573, and 586. When it finally burned out in the eighth century, a new power, Islam, arose to dominate much of the territory Justinian had ruled.

Germs: *Yersinia pestis*, which was discovered only in 1894 by Alexandre Yersin, is an especially large and complex bacterium. Since 2000, sequencing of human, bacterial, and viral genomes has allowed us to place these organisms on scales of size and complexity. *Yersinia*, first sequenced in 2001 by Parkhill and colleagues, like other bacteria has several thousand genes encoding enzymes and proteins, interacting in complex molecular webs whose functions are adaptively and rheostatically adjusted to different conditions. *Yersinia* is a living cell that survives by infecting the bodies of rodents, including prairie dogs, and is transmitted to humans via the unsuccessful biting efforts of infected fleas. *Yersinia* was not the only deadly bacterium of the time of the Black Death. Many deaths attributed to that pandemic may have been caused by other pathogens, including typhus, anthrax, and smallpox. All of these enemies of humanity were invisible and unknown until their discovery in the late 19th century, or, in the case of viruses, much later because of their very small size and difficulties in growing them. Later it would be learned that they were members of organismal types, bacteria or virus, radically different from one another in nature. Today, *Yersinia pestis*, the Black Death, lives on in many parts of the world, including North America. The disease smolders at a low level, infecting wildlife and a few people every year.

Yersinia, cholera, and many other pathogenic germs are bacteria. All bacteria are exemplars of the cellular theory of life: little, free-living agents that can do great harm when growing unchecked within our bodies. Bacteria are small by human standards, even compared to an ordinary human cell, as seen in the drawing below. However, viruses are usually much smaller, and have far smaller genomes and complements of genes, as Massung, Venter and colleagues found when they sequenced the genome of the smallpox virus (*Variola major*) in 1994, right at the dawn of the genomics revolution. The smallpox genome is only 186 thousand base pairs (kb), into which are packed 187 genes, compared to the 4000 or so genes in bacteria and the 25,000 or more genes in the human genome. The genomes of many other viruses are smaller still. The genome of SARS-CoV-2, which causes COVID-19, is tiny, like other coronaviruses, all of which have RNA genomes. Genome size is a notoriously poor indicator of complexity. Many organisms have genomes twice the size of humans, and small genomes of bacteria are packed tight with genes. However, the number of genes is a more informative index. Gene networks of cells often comprise hundreds, and even thousands, of interacting components, each representing a lever by which cell function and interaction can be calibrated, unfolding in developmental programs controlled by genetic switches and enabling cells to specialize to new roles.

Not so viruses, and especially simple viruses such as the virus causing COVID-19. When SARS-CoV-2 was rapidly sequenced following the onset of the pandemic in Wuhan, it was found to be only 30 kb, and only 15 genes, as shown below. Prions, misfolded proteins that are infectious because they can trigger normal precursor proteins to

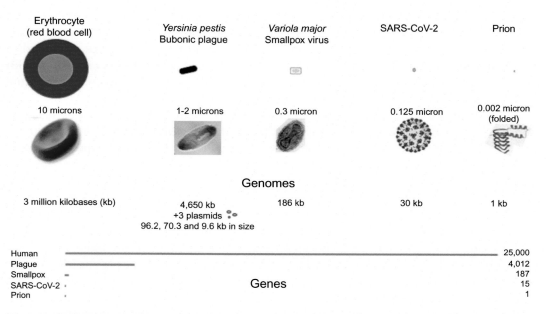

Erythrocyte (red blood cell)	*Yersinia pestis* Bubonic plague	*Variola major* Smallpox virus	SARS-CoV-2	Prion
10 microns	1-2 microns	0.3 micron	0.125 micron	0.002 micron (folded)

Genomes

3 million kilobases (kb)	4,650 kb +3 plasmids 96.2, 70.3 and 9.6 kb in size	186 kb	30 kb	1 kb

Genes

Human	25,000
Plague	4,012
Smallpox	187
SARS-CoV-2	15
Prion	1

The scale of cellular and viral life. One micron is a millionth of a meter. The smallest object resolvable by the naked eye is about 100 μm, 10 times as large as a red blood cell. *Created with BioRender.com.*

misfold, are still smaller. Each prion is only a relatively small protein, here its size expressed in terms of equivalent in DNA, as well as the micron, which is one millionth of a meter.

A virus executes its deadly program by using its limited repertoire of genes and proteins to enter and take over the machinery of a host cell. Having done so, the virus reprograms the cell to make enormous amounts of the components required to assemble new viruses. The multimeric viral coat, or capsid, spontaneously assembles from protein monomers that the host cell synthesizes on behalf of the virus. The coat protein shields the delicate RNA or DNA genome of the virus, and with varying degrees of specificity enables the virus to attach to cells. Coat proteins are sticky for certain cells, and natural selection refines this stickiness to increase infectivity and optimizes the machinery by which the virus takes over cells of specific species and types. Simultaneously, but usually far more slowly, the host evolves defenses.

Not everyone is equally susceptible to a virus. In 1996, when treatments for AIDS were in early stages of development, my friend and colleague Mike Dean discovered that people carrying an inactivating deletion variant of the *CCR5* (C-C Chemokine Receptor 5) gene were highly resistant to HIV infection and thus AIDS. If the virus cannot attach it is unable to infect, giving a small percentage of the population natural immunity. Mysteriously, the deletion variant is most common in European populations, as high as one in six copies of the *CCR5* gene in Finns, but much less common in people of African or Asian ancestry. Not enough is known about the biology of smallpox, but it has been hypothesized that smallpox, which killed so many Europeans, selected for the *CCR5* deletion variant.

Another possibility, introduced in a study by Stephens, Dean and others (in which I was a minor participant) (Stephens et al., 1998), suggested that the selective force was a bacterial disease, bubonic plague. Dating the *CCR5* deletion, it appears to have expanded in European populations beginning about 1000 years ago, a timing that is consistent with several of the pandemics, caused by different bacterial and viral germs, from the time of Justinian to the present.

Knowledge of the origins of viruses and their coat proteins that stick to target host proteins such as CCR5 is expanding all the time, but at the time of this writing at least 20 different types of viral coat protein monomers were known. What came first, the virus or the cell? Probably the cell, so that viruses could later hijack their machinery, but can we trace the processes by which viruses originated? As discovered by French microbiologist André Lwoff in 1950, many viruses can integrate into the host genome and sometimes excise later, enabling them to replicate, lyse the host cell, and trigger new infections. This observation led directly to the speculation that viruses originated from some host cell's DNA in the first place. We do not know, but recent advances indicate multiple, different origins for viruses.

As observed by Krupovic and Krupitsky, there are three main theories for how viruses originated in the first place, and these theories are not mutually exclusive. First viruses, and especially viruses with RNA genomes, may represent some of Earth's primordial life forms, the viral RNA encoding a small number of proteins and being protected by that small number of proteins. Arguing against this hypothesis is the fact that few RNA viruses infect archaea and bacteria. Instead, RNA viruses are mainly parasites of more evolutionarily recent, and derived, eukaryotes.

Second, and as seems more credible for a complex virus such as smallpox (*Variola major*), which has 187 protein genes, viruses may have degenerated from cells, gradually losing genetic elements nonessential to a parasitic existence. Far larger than smallpox are *giruses* (Van Etten, 2011). Giruses are recently discovered and as yet poorly understood giants of the virus world, and include viruses in the *Mimivirus, Mamavirus,* and *Megavirus* families. Giruses have not so much blurred as obliterated the dividing line between bacterial cell and virus. Although giruses are best classified as viruses and show gene similarities to poxviruses, giruses make the most compelling case for an ancient cellular origin of some viruses by a process of regression. Giruses can have more than 1000 genes, including some genes that are not derived from any cell in an obvious way, and their genomes measure as large as most bacteria and larger than some.

The genome of the smallest free-living bacterium, *Mycoplasma genitalium*, encodes only 470 proteins. Unlike simplistic genomes of viruses, the large complement of genes carried by giruses encodes many cellular functions. Also connecting them to cells from which they may have evolved by genomic reduction, giral genomes are large, hundreds of kilobases or in some species more than a megabase, and are composed of DNA, not RNA. Life may have begun as RNA, but cellular genomes are DNA-based.

Third, viruses may have begun within cells but as pieces of selfish DNA, such as the self-replicating transposons discussed earlier in this book, and they may have gradually, largely from host cells, acquired an armamentarium of protein genes that facilitated their infectivity. How would viruses have acquired these genes? Host cells make hundreds of types of proteins that enable cells to recognize and stick to each other, and a virus that integrated into a host genome can carry a host gene with it when the virus is excised and thereafter replicates. Intriguingly, some cellular proteins, for example tumor necrosis factor, have the ability to assemble into virion-like capsids. An important step in understanding the evolution of viruses would be to discover if the genes encoding these coat proteins were originally hijacked from a host cell. Krupovic and Koonin, analyzing the coevolution of host species and virus, which are prey and parasite in relationship, found that many and perhaps all viral capsid proteins were hijacked from cellular host ancestors, and sometimes multiple times. The molecular similarities are not high—even with the high mutation rates in viruses, especially viruses with RNA genomes, these viral gene thefts apparently occurred long ago. However, this makes sense—viruses are ancient and a virus probably only seldom can benefit from switching out its old capsid for a new one. In contrast, the replication genes of viruses apparently have no counterpart in cells, suggesting that they could not have been hijacked from any modern host but have origin either in some ancient, now extinct, cell, or that viruses are descendants of a primordial, virus-like lifeform.

As alien as viruses are to us, they are usually far more alien to each other. Nobel laureate David Baltimore identified seven classes of viruses, most of which are distinguished by a fundamental difference in their genetic code. Arguably, several could be classified as a separate kingdom or queendom of life. The genomes of humans, vertebrates, invertebrates, protozoans, and indeed all eukaryotic life on which viruses prey is based on double-stranded DNA—the double helix having been solved by Watson and Crick, using Rosalind Franklin's X-ray crystallography data. The genomes of Class 1 viruses such as herpesvirus and adenoviruses are most similar to ours because they are also based on double-stranded DNA. Class 1 viruses must replicate in the nucleus, where the host's DNA polymerase is found, and mainly can replicate in host cells that are actively dividing when the machinery of DNA replication is active. Class 2 viruses, including parvoviruses that infect dogs and were once proposed to cause chronic fatigue syndrome in people, have single-strand DNA genomes. Class 2 viral DNA replicates in a radically different way, as a rolling circle, but again in the host cell's nucleus.

Next, we consider RNA viruses. Bizarrely, Class 3 viruses are based on double-stranded RNA, and these viruses replicate in the host cell's cytoplasm, largely using the virus's own enzymes. Several Class 3 viruses infect people, including the rotavirus that has caused hundreds of cases of gastrointestinal disease among passengers and crew on a single cruise ship.

With Class 4 viruses, which includes the coronaviruses and hepatitis C, the picture is even more alien. These viral genomes consist of single-stranded RNA and their RNA genetic codes are in positive sense, such that the RNA can be, and is, directly translated by the host cell's ribosomes into the virus's proteins. If one tried to imagine what the most primitive combination of nucleic acid and protein might have looked look like, Class 4 viruses might come close.

Class 5 viruses, with their single-stranded but negative sense (antisense) RNA genome, are as primitive in their genetic code. At some point in the generation of new virions (viral particles), an RNA virus must replicate its genome. Like Class 4 viruses, Class 5 RNA viruses also do this in the host cell's cytoplasm, with the viral RNA polymerase first generating single-strand positive-sense RNA that can be translated by the host cell's ribosomes into proteins needed to assemble new virions. The Class 5 viruses have only been observed to infect animals, but are well adapted for that. They include germs that cause influenza, mumps, measles, rabies, and ebola, and several other types of distinct respiratory diseases in humans, and they cause canine distemper, bovine rinderpest, and a range of other diseases in other animals, sometimes crossing species boundaries from animal to human (zoonotic transmission).

Class 6 viruses, Baltimore's penultimate class, are, like Class 4 viruses, also positive-sense single-strand RNA viruses. However, their life cycle is fundamentally different. To replicate they must first be reverse transcribed into DNA and spliced into the host cell's genome. Thus they are known as retroviruses. Once integrated into the host cell's genome, the viral DNA is transcribed into RNA, and then translated into viral proteins for assembly of new virions. Viruses of this remarkable class include HIV, the cause of acquired immunodeficiency disorder (AIDS). It can be readily appreciated that a retrovirus, having integrated into the host genome, may cause infections that are long-lasting and that, while suppressible, may be difficult to extirpate without killing those host cells.

Last but not least, and showing that the dividing line between viruses of different classes may not be so bright, are the Class 7 viruses. Like the Class 1 viruses, these are also double-stranded DNA viruses, but unlike those viruses they "choose" to replicate through a single-strand RNA intermediate, and then that RNA is reverse transcribed back into a double-stranded DNA viral genome. Furthermore, and emphasizing the unity of life, the amino acid codes of all life on earth, as would later be demonstrated for these different viral classes, are almost identical to ours, recognized almost immediately after Nobel laureate Marshall Nirenberg deciphered the first "word" encoding an amino acid in 1963. The same RNA triplet UUU (uracil-uracil-uracil) that encodes the amino acid phenylalanine in humans encodes phenylalanine in *Ebolavirus*.

In this chapter and others, we have seen the Byzantine variety of individual self-replicating molecules such as prions and transposons and collections of self-replicating molecules such as viruses, giruses, bacteria, eukaryotic cells, and finally multicellular (metazoan) life. Organic life comes in all shapes, sizes, and types. Some is DNA based, some relies on RNA and, if prions are alive, some are protein. Host-parasite relationships such as the relationship of viruses to us, horizontal gene transfer across species hundreds of millions of years divergent, and transition from commensality to integration as occurred with the adoption of mitochondria and chloroplasts establish that life, including human life, is not definable in any strict qualitative way or on any quantitative scale. Human life is part of a phylogenetically connected and still horizontally interconnected webwork of genes and proteins extending across all these molecular forms. And yet, as will be seen in the next chapter, this pantheistic, pansyncretic view of organic life goes too far, blurring, if not quite obliterating, biologically and therefore philosophically meaningful distinctions. Our genomes, in executing their molecular computations, behave each and every day as if they "know" that they are different not only from the genomes of other species, but also that we are relatively more alien to even the unrelated members of our own species than we are to ourselves and our closest kin.

11

Altruism, of cell and self

Most human beings have an almost infinite capacity for taking things for granted.
Aldous Huxley, Brave New World

On a microscale, why does any cell of the body "gladly" sacrifice itself for the whole? On a macroscale, free-living organisms such as people and other animals often behave altruistically, but when they do, are they truly altruistic, or are they genetically programmed to do so?

We begin with the more difficult question of why some people and other free-living organisms sometimes sacrifice themselves on behalf of others at a moment of crisis. The answer remains controversial. A honeybee that dies as a consequence of stinging an intruder is not altruistic in a moral sense, because it is not even aware that it will die. The soldier who falls on a grenade to save his comrades has made an altruistic decision, influenced by training and innate predisposition towards altruism. However, neurogenetic determinism dismisses any moral subtext in the sacrifice of the soldier. Their death is no more morally significant than the self-sacrificing bee, both being governed by chains of causality. If a person who believes in neurogenetic determinism pins a Medal of Honor, Iron Cross, Silver Star, or Purple Heart on someone's chest, it is an artifice calculated to encourage valorous behavior.

John Eccles, a neuroscientist, attributed to God the ability of the human brain to produce altruistic behavior (Eccles, 1953). However, a god may have made the rules, but mathematics of inclusive fitness and natural selection favoring certain gene variants and gene combinations explain altruism. Eccles's God is not needed at the controls of the process, as explained by Donald Pfaff in *The Altruistic Brain* (2014). More interesting is the assertion that altruism is a quality that is a uniquely human attribute (Ramachandran, 2012). Is Ramachandran right? Many other species exhibit altruistic behaviors, and the eusocial insects in particular far exceed humanity in altruistic behaviors. However, while the genetic imperative of inclusive fitness may be shared between humans and bees, the path to the altruistic behavior—instinctive and implicit (unconscious) for bees versus instinct-influenced and partly explicit (conscious) for humans, is not morally equivalent. Even as they become more successful in manipulating behavior, neuroscientists remain far from understanding the origin of any complex behavior, but the evidence so far is that Ramachandran is more right than wrong.

Humans vary in their predisposition or capacity to make altruistic choices. They are unequal, both innately and by experience. The bee's behavior can be reliably predicted,

Immortal. https://doi.org/10.1016/B978-0-323-85692-8.00011-3

but not necessarily the person's. Why do some people give their lives in service to the community: the Hindu Dalit cleaning waste others will not touch, the nurse or nun devoting themselves to the care of the sick and dying? Sometimes, as with the lower-caste Dalit, because of birth and social constraint. Sometimes, as with the nurse, because of a more complex interplay between predisposition, societal norms, and choice. And on the other hand, why are people also ruthlessly competitive, and some more than others?

The subtitle of Pfaff's book was "How we are naturally good." However, biological mechanisms such as inclusive fitness do not make people naturally good. For example, inclusive fitness that can cause a male lion to risk its life for the pride equally well leads to infanticide. If the lion's valiant defense of the pride is "good" and the infanticide is "bad," this is mainly because we are anthropomorphizing—imposing a human moral construct on lion behavior. The lion who killed the cub was not a "bad lion." It was simply doing what lions do when faced with that situation. Infanticide is observed in about half of mammalian species studied, and can dramatically increase the chances that a male can pass on his genes by mating with a female who is no longer "wasting" resources on another's offspring. Humans and other species that behave altruistically at times are naturally good and cooperative at those times, but naturally competitive, unco-operative, warlike, and bloodthirsty when the circumstances warrant. If humans consis-tently treat stepchildren well, it is partly because this may be advantageous, for example to enhance the bond to mate and form community, in line with Pfaff's "naturally good." However, it is also culturally encoded and it is very much the consequence of moral choice. Resuming the comparison between humans and lions, "natural good" is nonex-istent. If a person saves a life out of instinct or because they were programmed by cul-ture, it may be expedient to call them "good," but actually they have only done what the lion might do. People who are "good" are the ones who are doing things we like because of explicit choice, and they may actually be naturally inclined to do otherwise. For peo-ple, who have a much wider range of choices and who can self-guide their own devel-opment (as discussed in my book *Our Genes, Our Choices*), more than instinct is required and fortunately evolution has given us the cognitive and emotional resources to deliver it. This is why people can be good, but in the end it is a choice, and for most of us it is one choice after another. Those choices should not be diminished by labeling them "natural," as in fact altruistic behavior in people is often quite unnatural and even fantastical, unless we understand that it arises from the decision-making of a moral mind.

The world is full of war, hate, and suffering. Against this disharmony, cooperativity is an aspirational goal. However, viewing the world's uncooperativity, can humankind greatly improve itself by emulating eusocial species that more faithfully follow the genetic rules of inclusive fitness? Altruistic behavior has been observed in many species, but it is within eusocial insects that altruism is most complete. Eusociality is not rare, and it is a highly successful evolutionary adaptation. Thousands of species of insects are eusocial, including ants (*Formicidae*), termites, wasps (*Apocrita*), and bees (clade *Anthophila*, in seven to nine families). Eusocial insects dominate the terrestrial ecology of all the world's continents except Antarctica. Ants alone constitute one-fifth of Earth's animal biomass. The ecological dominance of eusocial insects is due to the complex, eusocial caste structures of their societies and the self-sacrifice of individuals within them.

Ants, bees, and wasps (together, the *Apoides*), share common ancestry. Proving that eusociality was not inevitable, ants and bees first appeared between 110 and 130 million years ago, coincident with the rise of flowering plants. If Earth had been obliterated before then, and at any point during its 5-billion-year history, eusociality would not have arisen on Earth. Today, but not 200 million years ago, eusocial insect species are ecological keystones, a vexing problem because most insect species are in decline, and many are becoming extinct. They are key to the survival of flowering plants, with which they intimately coevolved. Each of 800 fig tree species is pollinated by a specific chalcid wasp.

It would be futile, largely pointless, and difficult for any person to name all species of ant, bee, and wasp. For example, there are at least 100,000 species of wasps in the families *Braconidae* and *Ichneumonidae*, almost all of which are parasitic, preying on eggs and larvae of other insect species with which they are coevolving. Ants and bees number more than 20,000 known species. By way of comparison, hominids (great apes) have existed for 25 million years, *Homo sapiens* diverged from its closest extant hominid relative, the chimpanzee, only about 4 million years ago, and the extant species of hominids are easily stored in human memory, with seven species in four genera: *Homo sapiens*, *Gorilla gorilla*, *Gorilla beringei*, *Pongo pygmaes*, *Pongo abelii*, *Pan paniscus*, *Pan troglodytes*.

Fortunately, it is unnecessary to intensively study or discuss the behavior of all these thousands of species to understand eusociality. This is not to trivialize the significance of any species or diminish the importance of its loss. For example, the extinction of an unrecognized wasp species can lead to the loss of a plant species that was key to an ecosystem. However, by the time one finished the recitation of the species of Apoides wasps, a new one might have been discovered, and, indeed nowadays, another might have gone extinct. In geologic time, which is not to be confused with human time, Apoides and many other species are rapidly evolving, and if humans wipe themselves out in the next few thousand years, there is a good chance that life on earth will be diverse and vital a million years hence. To discuss eusocial behavior, it is sufficient to refer to a few species as examples of biological solutions to ecological challenges and as representatives of genera, families, and orders: the domestic honeybee *Apis mellifera*, the fire ant *Solenopsis invicta*, the bald-faced hornet *Dolichovespula maculata*, to name a few that gardeners regularly encounter.

Advocates of eusociality can point to its endurance in insects from ancient evolutionary origins, and also to convergent evolution that many times has produced eusociality. The termite's ancestor—which is the always popular cockroach – diverged from the ant and bee some 470 million years ago, and eusociality evolved independently, and as will be seen, is maintained via distinctly different mechanisms. From fossils, eusocial bees of the same family (Apidae) as the modern honeybees are known from at least 87 million years ago. Within ants, bees, wasps, and termites, many species have lesser degrees of eusociality. Furthermore, eusociality has been lost and regained many times. The vast majority of wasp species are solitary, some of the solitary species building communal nests but not sharing division of labor or exhibiting other eusocial behaviors. However, the advantages of eusocial, cooperative behavior are often decisive, constituting a constant forward pressure of natural selection wherever there is an evolutionary niche that accommodates a large mass of individuals. A mature colony of bald-faced hornets contains 400–700 workers that hunt food and nurture larvae and construct their hanging nest, and many of them are always on hand to attack anyone who would dare disturb the nest.

Even so, the number and ferocity of hornets is dwarfed by several species of army ants. Colonies of army ants of many species and representing six subfamilies alternate between stationary and nomadic (legionary) phases. In nomadic phase the colony, which often numbers 15 million, consumes half a million other animals every day, many army ant species concentrating on robbing nests of other ants and wasps, the eusocial consuming the eusocial. On pheromone trails, inbound ants carrying prey occupy the middle lane. The ant trails are up to 20 m wide and 100 m long, and all around other animals have been flushed from their hiding places and are preyed upon by kleptoparasitic species, happy to take advantage of "the movable feast." As many as 500 other species, including different species of "antbirds," beetles, and mites, may follow one army ant species, *Eciton burchellii* (Rettenmeyer, 2011). Each night the army ants bivouac at a new site in interlocked, architecturally complex, living nests until the larvae spin pupal cases. Then they remain at one bivouac for about 3 weeks, feeding the egg-bloated queen and guarding her and the pupae. Simultaneously, millions of larvae and mature ants emerge (pupal eclosion), and again the army ants are on the move. Ordinarily, and if all has gone well, army ant colonies fission every few years, the old queen sometimes leaving. This event follows a longer bivouac required by the large size of the sexual male larvae. The worker ants affiliate with a queen based on her pheromone signature. When the queen dies, the workers often do not give up and die. They can fuse with another, possibly related, colony, thus increasing their inclusive fitness. Legionary behavior is highly successful and ancient and has evolved more than once. A genetic study of 30 species representing three army ant subfamilies and three continents—South America: *Ecitoninae*, Africa: *Dorylinae*, and Asia: *Aenictinae*—revealed that they all had a common ancestor (Brady, 2003) before the breakup of the supercontinent Gondwana some 100 million years ago.

Termites number over 3100 species, with many remaining to be discovered. Humankind is rightly credited with releasing carbon dioxide, leading to global warming, but termites are mainly responsible for the recycling of woody detritus into water and carbon dioxide. Dampwood termites (*Zootermopsis*) are among the most primitive living termites, helping to reveal how termite eusociality evolved. As Barbara Thorne and her colleagues at the University of Maryland have shown (Haverty and Thorne, 1989), one factor leading to their eusociality is intraspecies aggression—termite wars. It is unlikely that any individual termite will survive the aggressive depredations of other termites. It takes an army. Sometimes, hundreds of different dampwood termite families, each with its own queen and king, meet in the same fallen log, and when they meet, they battle. If the king or queen of the colony should die, any of the worker termites can themselves be promoted to king or queen, their gonadal development freed of suppressing pheromones. Unlike bees and ants, but like colonial wasps and hornets, termites are diploid and come in both sexes. We can see that genetically they are not so far removed from their cockroach ancestors. By their genetic and phylogenetic distinctiveness from ants, bees, and wasps, and by independent origins of eusociality in different termite species, termites illustrate that natural selection can repeatedly drive the evolution of eusociality in the world as we know it and that it has existed over tens of millions of years.

Altruism and false altruism: Darwin identified the self-sacrificial behavior of eusocial insects as posing the greatest challenge to his theory of natural selection. To Darwin, altruism was a riddle. The solution was first decoded, decades after Darwin's death, by JBS Haldane. Over drinks in a pub, Haldane declared that he would jump into a river to save two of his brothers or eight of his cousins, the degree of this geneticist's altruism being directly equivalent to his own genes that he would be saving by his sacrifice. Whether Haldane would have actually sacrificed himself in this fashion was never tested. Only in 1964 did William Hamilton and George Price express Haldane's idea of "inclusive fitness" in rigorous mathematical terms, their equation proving that an individual could optimize its reproductive fitness by sacrificing itself for the appropriate number of blood relatives, at a close degree of genetic relationship, or kinship. Hamilton's rule is an equation.

$$rnb - c > 0$$

$r =$ coefficient of relationship
$n =$ number of relatives who benefit
$b =$ benefit received by each recipient
$c =$ cost suffered by the donor

Hamilton's equation is deceptively simple in that two of its elements, benefit received by each recipient and cost suffered by the donor, are difficult to estimate. Just how much benefit (b) does an army of 15 million receive from the sacrifice of one soldier? On the other hand, in a time of war when death is likely at any moment, might the cost (c) actually not be so consequential? Does an individual truly sacrifice itself if it was likely to die anyway? And of course a worker ant cannot reproduce, so their only evolutionary value is to the hive. However, two of the parameters, coefficient of relationship and number of relatives who might benefit, are readily quantified.

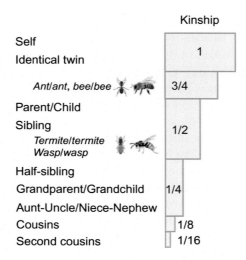

The coefficient of relationship (kinship) of eusocial ants and bees in colonies is ¾, intermediate between human identical twin and parent/child or sibling relationships. *Created with BioRender.com.*

Based on an understanding of genetic relatedness, we recognize that the altruistic behavior of eusocial insects is not altruistic in a cognitive sense (or whatever is going on in the mind of the insect) but is driven by genes. Their altruistic behavior is based entirely in the selfish imperative of genes to transmit themselves, very much in tune with Dawkins's selfish gene. Because termite kings and queens are primarily monogamous, worker termites in the colony share a 50% degree of genetic relationship both to their parents and to each other. Should one of their siblings be promoted to king or queen and have 1000 children, that will be the equivalent of themselves having 500 offspring. This is very impressive, considering that individually no termite is likely to survive. This evolutionary advantage favoring dissemination of their genes is equally persistent should they "altruistically" sacrifice themselves. In this way it should be evident that altruism, in the human sense of self-sacrifice to benefit others, is nonexistent in eusocial termites. They can accurately be viewed as little devices programmed to help the queen pass on their genes, and they are doing no more or no less than is essential for that to happen. Indeed, it is not unfair to compare the altruism or individuality of an individual in a eusocial society to that of a single cell in the body, those cells sharing nearly 100% of their DNA code with other cells in the body. A red blood cell also does not worry about its fate.

As shown in the figure, other eusocial insect species have a similar genetic compulsion to altruistic behavior. Hamilton's insight about inclusive fitness has very far-reaching implications in human social biology, some of which are at this point speculative.

Endogamy is a key prediction of inclusive fitness. It is usual for people to practice *endogamy*, associating and mating with individuals who are blood relatives at cousin, second-cousin and third-cousin degrees of relationship, even as incest taboos prevent genetically riskier consanguineous matings between first-degree relatives (brother-sister, parent-child) and, usually, second-degree relatives (for example, aunt-nephew, uncle-niece).

In some situations *hypergamy*—mating with someone unrelated—is favored, facilitating the dissemination of *genes* and reducing inbreeding and the impact of the genetic load of deleterious recessive alleles. In making the intriguing case that sexual equality in hunter-gatherer bands weakens endogamy, a group of anthropologists led by Dyble and Migliano (2015) collected data that powerfully proved the opposite, verifying endogamy and genetically predicted affiliation. The hunter-gatherer communities they studied, the Palana Agta and the BaYaka, were mobile, multimale, multifemale bands with enduring male-female pairbonds but extensive cross-parenting, subsisting by cooperative hunting, fishing, and gathering. These bands, most often numbering about 20 adults, are egalitarian, sharing resources freely, even with individuals who are genetically unrelated, as seemingly contradicted the inclusive fitness model. However, blood relationship predicted coresidence in the hunter-gatherer bands, and even more strongly predicted the multifamily composition of a primitive agricultural population. In the hunter-gatherer bands, only 1 in 6 individuals was unrelated to someone else in the band, and in a comparison group of "primitive" farmers, the Paranan, less than 1 in 20 were unrelated. Some tendency towards hypergamy was seen in that in these small bands, half of matings were not with blood relatives, although—viewed the other way—a 50% rate of consanguineous and affinal mating is a very high price to pay, given the penalty for autosomal recessive diseases that arise from such matings. Clearly, there is a strong impulse to affiliate, mate, and cooperate with blood relatives, even as incest taboos discourage matings between people who are most closely related.

Social behavior is governed by constellations of genes working in concert. Elsewhere, this book also discusses the role of constellations of genes, and their preservation, in speciation. In various species, gene constellations are preserved on large chromosomal segments known as supergenes. The human X and Y chromosomes can themselves be viewed as supergenes. Endogamy is also a way of preserving such constellations, but even occasional outbreeding can disrupt them, unless the multiple genes involved are on the same chromosomal segment, as happens with supergenes and the sex chromosome. Because the human species is substantially outbred (despite whatever best efforts are made to avoid breeding with "the other"), endogamy and genetically predicted affinity probably have more to do with the preservation of numbers of genes rather than their combinations. Certainly, inclusive fitness partly explains monogamy, pair bonding, and the enormous investment parents put into their children. It explains the tendency of closely related tribes to cooperate and even coalesce and, on the other hand, the tendency of larger tribal groups to fracture along lines dividing genetically related clans, for example as described among the Yanomamö people by controversial anthropologist Napoleon Chagnon, who among other things may have facilitated violence among that tribe (Chagnon, Yanomamö: The Fierce People, 1968).

Conjecturally, inclusive fitness may partly explain homosexuality, in which the individual's own fertility is reduced but perhaps with the benefit of enhancing the fertility of a parent or sibling, as discussed further in *Our Genes, Our Choices*. Also conjecturally, inclusive fitness can explain some aspects of the parental imprinting of certain genes such that the father tends to imprint these genes (such as fibroblast growth factor) in a way that will make the offspring larger and more vigorous, but also more demanding of the resources of a mother with whom he may or may not again mate.

Inclusive fitness probably explains man's continuing fascination with race and racial identifiers. In a small village, such identifiers are unnecessary because people know who is related to whom. In modern cities, in which few of our neighbors are related to us by blood, more than 90% of human genetic variation is interindividual and untrackable without DNA testing, but people have perhaps latched on to whatever superficial racial identifiers are discernible in order to track the small percentage of others' genes that are more likely to be shared because of population of origin. Conjecturally, inclusive fitness can explain the modern obsession with using pedigree and DNA analysis to identify other people with whom we share common ancestry, and thereby more genes than by random chance.

Recently, the simple, elegant theory of inclusive fitness was attacked. The attack was provocative and thought-provoking, although a failure. Martin Nowak, director of the Program for Evolutionary Dynamics at Harvard, and author of *SuperCooperators: Altruism, Evolution, and Why We Need Each Other to Succeed* (Nowak and Highfield, 2011), together with mathematical biologist Corina Tarnita and the founder and giant of sociobiology, E. O. Wilson, evaluated inclusive fitness against a variety of specific examples of eusocial behavior from insect species. Several years earlier, in a 2008 paper published in *Science*, Nowak had unified the mathematical basis of five types of social cooperation: group selection, kin selection, direct reciprocity, indirect reciprocity, and network reciprocity. In his analysis of inclusive fitness among eusocial insects, what Nowak claimed in a report published in 2011 in *Nature* is that the mathematics of inclusive fitness hardly ever worked, when reality-tested in this way. Nowak, Tarnita, and Wilson concluded that inclusive fitness was not necessary to understand the origins of eusocial behavior. However, Nowak's paper stimulated five critical letters in *Nature*, one with 137 signatories (Abbot et al., 2011) and other thoughtful criticism (Rousset and Lion, 2011), and when the dust settled, kinship selection emerged as strong as before.

There are several keys to why inclusive fitness stands as a force maintaining altruism and social structure of many species. Inclusive fitness is a special case of natural selection, taking into account the actions of genes on the fitness of other bodies. Therefore, the debate initiated by Wilson and company was one that from the beginning was overdramatized, if not completely unnecessary. This conclusion was foreshadowed by Nowak's own work on the mathematical bases of different types of social cooperation, and by Wilson's vision of eusocial insect colonies as superorganisms, as will soon be discussed.

Kinship selection (inclusive fitness) has been successful in explaining many individual behaviors, for example alarm calling, and social structures (castes of nonreproductive individuals, sex ratios of different species—there being thousands of papers on that subject) that otherwise would be unfathomable. Inclusive fitness explains parental care—e.g., why a male emperor penguin stands in a blizzard for weeks incubating a fertilized egg when it might be off somewhere else enjoying itself and eating fish. The fertilized egg carries 50% of the father's genes, but he can only be sure (if he thought about it, which he does not) of the genetic contribution of the mother. The father emperor penguin is genetically programmed to protect and care for the fetal penguin in its egg, not because he likes eggs but because by incubating his mate's egg, the penguin may thereby transmits his genes. If the chick survives, he is highly likely to transmit his genes because emperor penguins are monogamous.

Kinship selection, together with the details of social behavior, also explains the evolutionary origins of eusociality. In Hymenoptera, the insect order encompassing all wasps, bees, ants, and sawflies, and at least 130,000 species in total, eusociality has independently evolved many times. However, in each case, where the origin of the eusocial species can be traced, the queens in the ancestral lineage were monogamously mated (Hughes et al., 2008). The direct implication is that all that queen's offspring were related at the full-sibling level, 50%, and not the half-sibling level of 25%. The full-sibling relationship of all the queen's "children" fostered their cooperative behavior, a finding that makes perfect sense under inclusive fitness, but less sense if ants, bees, and wasps are for no particular reason predisposed to cooperate with other insects of more or less the same kind. They do not. In a log, termites of the same species fight wars. Under Nowak's cooperative behavior model, we would not be able to understand why. If it is in their nature to cooperate, why cannot they all get along? However, under the inclusive fitness model the answer is evident—they fight to make it more likely that their closest kin transmit their own genes.

Cooperation is a worthy virtue, worth encouraging, but when people who believe in it seek to make it a general virtue of nature, it becomes a buzzword, whether or not cooperation is spelled with an umlaut. Discussing cooperation as if it were programmed rather than taught and practiced destroys cooperativity as a virtue. Therefore it is important to recognize the evidence that species are cooperative and also viciously competitive. People are naturally both.

Dawkins, writing about Nowak's cooperation, said, "'Cooperation' is not a third principle on top of mutation and natural selection, it is a behavior that evolves by either natural selection (as it must have done in the many species that do cooperate without culture, like social insects) or is socially mandated by complex creatures like humans." Dawkins's conception of the origins of human cooperativity can perhaps be taken further, and in a slightly different direction. As many have observed, the key difference between the cooperative behavior of large groups of humans and large groups of eusocial insects is that our cooperativity never evolved for large social groups. However, our ancestors had to cooperate to survive and succeed in small social groups bound together by close genetic kinship. Thousands of generations of cooperation and sacrifice for the tribe left a residue of cooperativity essential to the coalescence of unrelated or at best distantly related individuals into large, partially cooperative, communities. The cooperativity can be socially encouraged, and for example military training and common experience lead soldiers in a platoon to feel that they are a "band of brothers." Cooperativity is then tested time and again when people encounter others who are alien, and who can therefore not trigger a glimmer of altruism. Cooperativity is continually challenged by resurgent "tribalism," instantiated in various forms, including identity politics, race-tinged gangs, overt racism, and these days, the iron pyrite of ancestry analysis.

In seeing nature as a cooperative, Nowak stepped beyond the realm of science and into that of faith that the compassionate biblical virtues are encoded in our genes: "The mathematical analysis shows that winning strategies in the game of co-operation have to be hopeful, generous and forgiving." Nowak is partly right, as just discussed, but he is mainly wrong. Said Dawkins, "…maybe altruism and compassion are in our genes, but so perhaps are aggression, spite, xenophobia, and hatred." The principle of cooperativity can explain why people get along, but fails to explain why they do not, and why they tend to get along with some better than others based on who they are, or who we think they are, and not how cooperative or deserving they are. In the end, the prodigal son is welcomed back because he is one of us and has a kinship coefficient of 50%.

The Cinderella effect: Kinship selection explains the otherwise mysterious Cinderella effect, identified by evolutionary psychologists Margo Wilson and Martin Daly at McMaster University. The Cinderella effect, memorialized in fairy tale but all too real for some children, may be an evolutionary echo or remnant of male infanticide in our ancestors. Among primates, males frequently kill the offspring of other males, thus bringing the mother back into estrous and freeing her, and him, from bearing the evolutionary cost of caring for a juvenile that does not carry his own genes. In some non-primate species, infanticide is normal. When a male lion takes over a pride, it is common for him to kill all cubs fathered by other males. Obviously, a human stepfather or stepmother cannot behave in this fashion. In prehistoric times, the female, and probably other members of the tribe and especially her blood relatives, would have rejected the murderous male. The strange male had to behave within bounds of social custom and propriety. In historic times the force of law has joined with spousal and social opprobrium to reduce the frequency of infanticide, and yet an echo of the Cinderella effect is seen in human families.

In most parts of the world, in prehistoric times, and as observed in Oceania (Silk), when people adopted they were more likely to adopt a child who was genetically related, for example at the aunt/uncle-nephew/niece level (25%), as often happened, or at the cousin (12.5%) level. On the other hand, foster parents are more likely to kill adoptive children despite having gone through the effort and screening involved in adoption.

The Cinderella effect is an unpleasant fact, although there are some important nuances that are seldom or never discussed, even by critics of the idea. Adopted children may be more likely to carry genetic loading for traits that provoke violence or rejection from foster parents. They may also have been more likely to have experienced intrauterine teratogenic exposures, including exposure to alcohol and other drugs and effects of poorer nutrition and prenatal care. The effects of such exposures and inherited genetic variation can be subtle, but important. Also, the child, mother, and father may not have benefited from early bonding experiences that occur during the first weeks and months of life. These possibilities, all of which make the child more than the passive object of the behavior of the adoptive parent or stepparent, remain to be investigated, but will require more fine-grained analysis, for example incorporating genomics, epigenetic measures of the imprint of environmental exposures on the genome, and neurofunctional studies on parents and their biological and adopted offspring, than the types of studies that have so far been performed. It has been proposed that adoptive parents are somehow more pre-disposed to violence against children; however, and as observed by Wilson and Daly, this is refuted by comparing the murders of children who are blood relatives versus adoptive children within the same family. This type of data is not extensive, adoptive parents as a group being loving and giving and having lower rates of child abuse and neglect than other parents. However, in two studies the adopted sibling was murdered in 9 of 10 instances in one and in 19 of 22 instances in the other. Of far greater social impact than the rare murders are the many instances in which the genetically related child is favored in parental attention, education, and even health care. The danger to the adopted child can be unintentional. Stepchildren are also at far higher risk for unintentional injuries, including accidental fatalities, as can reflect lower parental vigilance to danger (Tooley, 1972). In the absence of real data bearing on ad hoc explanations for the increased parental violence and decreased attention directed against adoptees and stepchildren, the most usual explanation of why a stepmother is an evil stepmother appears to lie in kinship selection.

Superorganisms: Based on their interactions and evolution, E. O. Wilson, at an earlier time before he attacked inclusive fitness, identified colonies of eusocial insects as superorganisms. The individual insects in a colony function as cells within a larger body—obviously a very powerful conception of inclusive fitness. The superorganism concept has been tested and confirmed in new ways. A team of scientists at the University of Florida led by James Gillooly (Hou et al., 2010) compared colonial insects to other animals in terms of metabolic scaling, which is based on rates of growth and reproduction, and lifespan. They analyzed 168 species of eusocial insects, including bees, wasps, ants, and termites, studying the metabolic scaling of whole colonies of these insects. Viewed as superorganisms, the rapid growth rates and long lifespans of colonial insects is in line with that of individual large multicellular animals. The queen may have a lifespan of several decades, as is otherwise unexpected in an insect.

Casteing call: Eusocial societies of insects illustrate that when individuality is dis-counted for the good of the superorganism, this extreme socialization drives both the sac-rifice and involuntary task specialization of the individual. Much as the body differentiates its cells into a myriad of specific types to build muscle, nerve, blood, and organs, so does the social community also follow the imperative of specialization. For people, as for a cell in the body, the question is never one of whether one will specialize, but only the process by which a person will find his role. Like the cells in the body, whose genetic program is differentially expressed due to epigenetic changes in DNA and chromatin enabling cells to differentiate, epigenetic change also underlies the ability of social insects to specialize into different castes.

Humans have repeatedly devised and evolved social structures for task specialization that emulate but that far exceed in their complexity and level of distinction and differen-tiation the behaviors of eusocial insects, including their caste systems. Every one of the complex social behaviors that have been observed in hymenoptera and termites has been seen in human cultures, the difference being that, in the main, humans have adopted these behaviors and, almost in their entirety, insects have adapted them.

The behaviors encoded into the brains of eusocial insects shows that gene constellations are polymorphous, and astonishing in their precision and complexity: slave-taking (e.g., the Amazon ant, *Polyergus*, a genus of 14 species, all of which are entirely dependent on workers captured from other species in frequent massive colony-to-colony raids), parasitism (e.g., the ant *Strumigenys xenos*, which is entirely dependent on food gathered by a related ant species whose nests it penetrates), burial of the dead (necrophory), tutoring (one ant or bee guiding others to food), dancing, nomadism (army and driver ants), commensalism (including the gut protozoans that enable termites to digest cellulose), cannibalism, excavation, gardening (leaf-cutter ant), care of the young, feeding of others, mimicry, rafting, gliding, husbandry of other animals (including aphids and caterpillars, depending on the species), predation, scavenging, war, and building of nests and cities.

However, people, in a few generations, create and then dissolve patterns of social behavior that outwit, outdo, and outperform what it takes thousands of generations, for natural selection to accomplish in insects. Whereas people engage in all of the behaviors listed in the preceding paragraph, a eusocial insect species, even with its caste system, has a limited repertoire. It has been stated that natural selection is also evident for human social organization, the efficient modes of social organization succeeding and leaving descendants and the inefficient systems perishing. However, the conception of human social systems as evolving is true only in a descriptive sense, as narrowly applied to certain situations.

Far more often, and in a way that is usually impossible for the genetically driven social evolution of insects and other animals, human social change occurs as revolutionary, salutatory, in jumps. A monarchy is overthrown and replaced by a democracy, as in the United States, or by a theocracy as in Iran, by a dictatorship as in Fascist Spain, or by a communist oligarchy as in the Soviet Union. Only a few generations later, there may be another social convulsion. The genes have not changed, but the people have transformed the way they live. Often under the leadership of individuals of great power and influence, the likes of which no eusocial insect species has ever seen, people can almost instantly, but often within a single generation, adapt and emulate the most complex social structures of government. This social change is not evolutionary in either the Darwinian, genetic, or Lamarckian, epigenetic, sense. This does not happen among eusocial insects. For them, social change requires genetic change, as may require the spread of a DNA sequence variant or combination of variants, or awaiting a new mutation.

There may be social stasis or slow change for many generations, and then when things suddenly change people learn that the most important lesson of history is "things change." Frequently in biological evolution, as distinguishes it from social change, there is no viable pathway by which a species can traverse the zone of extinction from one adaptive peak to another way of existence where it could be highly successful. Also, there is sometimes no reason for a species to change. The adaptive challenges of the coelacanth were so little altered that it has maintained its form, and probably most of its functions, over hundreds of millions of years, even if its molecules were steadily evolving as the molecular clock ticked over that time. For human cultural evolution, change is more the rule than the exception. Large-scale social change is risky because any single change often needs to be balanced by others, but many of a society's laws and customs can be changed simultaneously or in short order, potentially reestablishing a new equilibrium. The biggest risk to a society is stasis. As illustrated by America's Federalist papers and founding documents, even when a social system is mapped out beforehand, it is just the starting point. A bridge is engineered to last, and eventually to be replaced or kept as a historical relic. A fish may occupy an ecological niche for millions of years. The landscape in which a society competes constantly changes. Years after their founding, societies—like coelacanths—may look superficially similar and bear the same name, but like the coelacanth they are evolving, and far more rapidly, or they do not survive.

Conclusion—Morality of social systems: For other animals, there is no moral dimension to a social system or social behavior, even if that behavior is slavery or infanticide. Due to their dumb nature, the presence of a caste system poses no moral questions. Indeed, one could even argue that it would be unkind to remove a honey ant from its designated, and to a human way of thinking, bizarre role as a bloated nectar storage vessel. To many people who make their living from the land, any animal on their property is either livestock or vermin. To the social engineer designing the utopian society, there can also be a tendency to think of people in terms of classes or types, as has been implemented in hereditary monarchies, slave-holding countries, and countries with caste systems. In the modern societies where princes, slaves, and Dalits are no longer to be found, people are nevertheless sorted by race, gender, class, and place of birth. The classifications are useful for understanding what is going on, and at times to target resources to groups that need them to rectify inequity. The classifications may also mislead us to evaluate the behavior of the individual from the perspective of class, or to ignore the individual entirely, focusing on class. Increasingly, reminiscent of 20th century eugenics, political opponents are claimed to be inherently different, and inferior.

12

Shrines and museums

A red light high in the nave of the Milan Duomo marks the location of a glass reliquary—a safe-box for a sacred object. Nearby is the Nivola, an ancient word for cloud, said to have been designed by Leonardo. The Nivola is an elevator operated by a clever system of winches, now electrically powered. Around September 14th of every year (the nearest Saturday), the Bishop of Milano ascends above the congregation to the reliquary. He descends to the altar, and repeats a ceremony initiated in 1576 by Bishop Carlo Borromeo, who used the relics, wood fragments of the true cross and a nail used to crucify Jesus, to end an epidemic of bubonic plague. All who have benefited from the relics can thank the mother of Emperor Constantine, Helena, who discovered the artifacts in Jerusalem around 326 and gave them to her son.

Most renowned artifacts of our time and times past, from the Shroud of Turin to *Salvator Mundi*, a purported painting of Leonardo's, are, like the nails and wood of the cross, valuable because of their history rather than their rarity in nature or artistic composition. Nails, pieces of cloth, bits of wood, and even paintings of the quality of *Salvator Mundi* are commonplace. Their value is bound to provenance. About a decade ago, *Salvator Mundi* was acquired for $10,000 dollars—costly enough, given its shabby condition. It was then extensively restored and rediscovered as a Da Vinci. It may or may not be. Recently, it was sold at auction for $450 million to a Saudi Arabian prince.

Obsessive collecting is distinct from the reverence for things, but passion for things shades into reverence. The thing collected may be cached and hidden, like an acorn woodpecker or squirrel hoarding nuts, or displayed in special albums, cases, or airless chambers—or over fireplaces—as stamps, action figures, paintings, and totemic masks often are, or pawed over, as an illuminated medieval manuscript may be if it has fallen into the wrong hands (something I once witnessed). Trophy hunters satisfy a deep inner urge by collecting heads, horns, and skins of commonplace species, but are even more gratified to bag and display the rare and exotic. The utility of these collections is not obvious. Sometimes it lies in the magical, medicinal, or totemic power with which the object is supposedly imbued.

Clearly, the passion to collect goes beyond, and probably began, with the inanimate, whether originally derived from a living entity or not. Originally it was probably derived from tool use and mating ritual. Before they acquire speech, children, as well as the young of nonhuman primates and other tool-using species, for example crows, pick up, carry, and store objects. At the age of 3, my oldest son Aaron was inconsolable if unable to collect

Immortal. https://doi.org/10.1016/B978-0-323-85692-8.00012-5

Salvator Mundi, attributed to Leonardo da Vinci.

and carry every stick shaped like the number seven, which complicates a walk in the woods. My daughter Evir insisted on collecting all perfectly ovoid rocks, which entirely changed or prevented a stroll along a cobble beach.

For humans and other species, the transactional demands of social and sexual success require the avid collection of objects that are useless as tools. And thereby are we instinctively programmed. Nowadays, but far into the past and in no way different from bowerbirds, males offer precious objects to females, and often such gifts are given by females to males, and exchanged between partners of the same gender to enhance the pair-bond. The modern equivalents of shiny pebbles, exotic shells, and feathers include red sports cars, silver, gold, diamonds, bitcoins, and cold hard cash. People, in turn, enhance their beauty or handsomeness by displaying plumes of birds on their heads and precious metals on their bodies.

The craving to collect extends to a preference for the original, authentic, and rare over the common or the identical copy. This is the difference between gold and gold plate or between a ring with a purely decorative flawless diamond of a high grade versus a slightly flawed diamond costing half as much. The point is not whether you can tell the difference but whether an expert can, after careful study. The world's most expensive stamp, at $2.3 million, may be the subtly intriguing 3 skilling Swedish banco (1855), printed in the wrong color, rather than blatantly bizarre inverts such as the inverted U.S. Airmail Curtis Jenny (1918). In good condition, the original 24 cent stamps are worth >$1 million apiece.

A souvenir version of the inverted Jenny released in 2013, and worth a few dollars. The most famous stamp forger was Jean de Sperati, who was finally paid off by the British Philatelic Association in 1954 to keep his engraving tools out of others' hands.

Even when the real thing is neither sacred nor valuable in any utilitarian way, people will pay almost any price for authenticity and specialness. Around the beginning of the 20th century, tobacco companies hit on the idea of putting baseball cards in cigarette packs. However, Honus Wagner, one of the first five players to be elected to the Baseball Hall of Fame, and already revered in his own time, did not approve of his image being used to promote smoking, as the weight of the evidence suggested. Or, as some maintain, Wagner may have wanted more money. Only 50–200 of the cards were released before Wagner put a stop to their distribution. Depending on quality and details of provenance, one can purchase a T206 American Tobacco Company Honus Wagner (or forgery), for hundreds of thousands or millions of dollars.

Collectors' preference for the original baseball card or stamp is unlike reverence for authentic religious relics. Certain high-profile stamp collections even focus specifically on forgeries. Whether or not the Nail of the Duomo or the Shroud of Turin were physically associated with Jesus, they would be precious artifacts of a time long past, but they are not revered as relics because they are ancient. The Dead Sea Scrolls are valued both for their antiquity and meaning and not because anyone thinks they were written by a divine hand, as many believe the tablets of the Ten Commandments or the Golden Plates of the Angel Moroni were. The Dead Sea Scrolls are precious for scholarly analysis and as remnants from which the original text and meaning might be most accurately reconstructed. The Scrolls and other archeological artifacts can also validate the faith of the living in long-ago events. In a lesser way, a postage stamp or baseball card can connect us to the past, but not with the sacred. The centuries-old searches for the Ark of the Covenant, Holy Grail, fragments of the True Cross, and Ark of Noah are motivated in part by the mystical powers that could be unleashed and put to use if ever we could lay hands on them. We may have a scholarly and historical interest in the Dead Sea Scrolls, or be charmed or obsessed by stamps or old photographs, but do not impute mystical powers to them.

From antiquity, people have preserved sacred objects purported to have magical or divine properties: the Egg of Leda by the Spartans, or the oracular head of Orpheus kept at Lesbos. Buddhist stupas hold the ashes of the prominent, or even body parts of Buddha himself, the tooth of Buddha having made its way to Sri Lanka. In Christianity, the veneration of relics is as widespread and conventional as to discourage comment. By 787 the Second Council of Nicaea declared that the altar of every church should contain a relic, and the Council of Trent (1563) instructed that the remains of martyrs, though not worshipped as things unto themselves, should be revered "for through these many benefits are bestowed by God…."

To this day millions of pilgrims journey to sites holding relics and to the Holy Lands of Israel and Saudi Arabia, making these pilgrimages and hajjes both from religious duty and for spiritual and physical healing. The purpose here cannot be to exhaustively list and classify relics or the pilgrimages made to sites holding them—however entertaining and sometimes moving as these stories can be—but to distinguish relics from mere collectibles and pilgrimages from travels to collector's conventions or museums where collections are held. As will soon be discussed, the world's great museums hold a vast variety and number of invaluable, and less valuable, human-derived artifacts. However, some museum artifacts, for example, the Staff of Moses, a hair of Muḥammad's beard, the turban of Joseph, and the sword of David at the Topkapı Palace museum in Istanbul, or a *cuauhxicalli*, a sacrificial bowl used to hold human hearts for the Aztec blood god sometimes on display at the Los Angeles County Museum of Natural History, are meaningful in an entirely different way. Confronted with a bowl that was used in bloody, sacred ceremonies, however ancient, the boundary between the sacred and the secular becomes hard to discern.

To believers, the ashes, bodies, bones, and accoutrements of the holy are sacred, and therefore the guidelines for their display, if they are displayed or held at all, is bounded or guided by rules beyond those we would devise for the secular. The question that was first posed in the latter half of the 20th century is whether or not the holdings of physical remains of humans who did not happen to be kings, saints, or prophets are bounded by the same rules. Is a display of human skulls bounded by no different rules than we would apply to collections of paintings, baseball cards, or stamps? What is the value of Einstein's brain, a spear that pierced the side of Jesus, or a living cell derived from a celebrity?

13

Robbing the grave

The dead are all around us, except perhaps in some of the coldest, deepest, and highest places, or subterranean, within the interior of the earth. The gods, and those blessed to join them, supposedly lived on high mountains, clouds, or in the starry heavens, and other gods and the condemned lived in the underworld. But except under special circumstances—Orpheus, Dante, mountaineers, spelunkers, astronauts—the living could not visit these places.

Many cities of the living were built on the rubble, middens, and graves of the ancients (James Michener, *The Source*). The stones of ancient cities are repurposed and mortared with bones of the dead. People devote enormous space and wealth to cemeteries, necropolises, and shrines to the dead, and after a time these monuments are then themselves repurposed. Still, all this evidence of death that surrounds us is seemingly not enough. If the dead do not rest easily it is perhaps because their tombs are too often looted, often within years of burial, but also even hundreds of years later in the name of art or science. Their skeletons, mummies, and the artifacts with which they are interred are assiduously collected. Nineteenth-century Europeans invented the science of archeology, excavating tells of Asia Minor and searching out undefiled barrows and crypts for the treasures of the dead. On any beach is a person with a metal detector looking for treasure in the sands, and offshore treasure hunters scour the oceans for shipwrecks laden with precious metals and artifacts of value.

Until quite recently, archeology was not a very scientific science, the activity being more akin to socially sanctioned grave robbing, the emphasis being the industrial collection of artifacts to stock museums, and the hunt for valuable treasure. Indeed, the 18th and 19th centuries were a time when the collector's impulse, found in most people, was at its zenith. Under the cover of academic interest, ancient necropolises were looted of their most superficially attractive if not scientifically informative objects, the stratigraphy and orientation of their contents disturbed in ways that would prevent reconstruction of original relationships and meaning. The insatiable demand for precious metals and historically precious artifacts created an economy of poor and often poorly educated grave robbers, greedy middlemen, avid collectors, astute investors, hypocritical curators, and creative forgers. The value of gold was easily established. The authenticity of other items, or any item made more precious by provenance, was always in question, and justifiably so, as modern methods have shown. Technologies to unmask the fakes drastically advanced, but many of the artifacts that continue to be valued by museums and private collectors, and on which the careers of scholars and "curators" are built, are fake. Despite any evidence to the contrary, some will always believe that Jesus was wrapped in the Shroud

Immortal. https://doi.org/10.1016/B978-0-323-85692-8.00013-7

of Turin. In contrast to many other types of artifacts, skeletons are harder to counterfeit but, as the case of the Piltdown Man proves, bones can be mixed and matched to fool experts; perhaps a few of the most exciting paleoanthropologic "discoveries" of the 21st century will also be ultimately debunked.

Where have all the skeletons gone? There are hundreds of thousands of burial mounds across America, and many Indian tribes ceremonially buried their dead. Dotting Meso-America are abandoned cities—necropolises. In 1989 the Smithsonian Institution and other American museums agreed to return skeletons and accompanying artifacts and ornaments: beads, clothing, weapons, tools, and ceramics of thousands of American Indians. At that time, the Smithsonian had some 18,900 skeletal remains.

Suzan Harjo, then Executive Director of the National Congress of American Indians, said, "The fact that the Smithsonian has 19,000 of our people is one of the last vestiges of colonialism, dehumanization and racism against our people." Despite the fact that the remains had been scientifically studied, it was unclear how many people they represented. But as Harjo noted, the skeletons represented a lot of people: "Four thousand five hundred alone are from the Army Medical Museum from people whose heads were taken at the turn of the century for what they called the Indian Crania Study." One head, one person. The remains included some of Harjo's own ancestors, victims of the Sand Creek, Colorado massacre in 1864, after which the heads of some of the dead were lopped off and collected.

American museums have returned thousands of skeletons, but if you are worried that the museums are running low, stop worrying. In *Bone Rooms* (2016), Samuel Redman, an assistant professor of history at the University of Massachusetts, Amherst, noted that as of 2015 the National Park Service estimated that over 50,000 human remains had been repatriated under NAGPRA (the Native American Graves Protection and Repatriation Act) passed in 1990. The Smithsonian museums hold some 30,000 human remains, most of which will never be displayed and almost all of which have not been adequately studied to give up the secrets of the dead. Other museums also have thousands of skeletal remains, for example 18,000 at the Hearst Museum of Anthropology at the University of California, Berkeley. According to Redman, there are probably another half million Native American skeletons in European museums. It is estimated that half a million American Indian remains are still in American museums, plus tens of thousands of skeletons from Black Americans, European Americans, and indigenous peoples from all over the world. Ten thousand here, thirty thousand there—it adds up. Worldwide there are probably millions of human skeletons in the museums. The difficulty with repatriating these materials is that because of their nature, and the ways they were originally acquired, their cultural connection to any living people cannot be unambiguously determined.

Hoarding all of this stuff, little of which has ever been displayed or studied, had in the first place represented a massive enterprise. However, archeology, especially as practiced in modern times, should not be directly equated with hoarding. It would be wrong to categorically condemn or paint in crude brushstrokes this field of science that has made staggering contributions to our understanding of all aspects of human nature. However, the resemblance of 19th century archeology to hoarding, and at times the motivations, are more than superficial. In any case, most of the human remains were not collected by professional archeologists but were originally unearthed or removed by a variety of private donors including explorers, soldiers, physicians, people for whom collecting was a passion.

Obsession leads to a compulsion to collect and hoard a variety of inanimate and animate objects. In the Diagnostic and Statistical Manual of Mental Disorders (DSM-5) of the American Psychiatric Association, hoarding disorder nests like a rare and irresistible bird within a group of compulsive disorders, the criteria being:

1. *Persistent difficulty discarding or parting with possessions, regardless of their actual value;*
2. *A perceived need to save the items and distress associated with discarding them;*
3. *The accumulation of possessions that congest and clutter active living areas and substantially compromises their intended use. If living areas are uncluttered, it is only because of the interventions of third parties (e.g., family members, cleaners, or the authorities); and*
4. *Clinically significant distress or impairment in social, occupational, or other important areas of functioning (including maintaining an environment safe for oneself or others)*

Other diagnostic specifiers of hoarding disorder include whether the individual shows excessive acquisition of unneeded items in addition to difficulty discarding them and lack of insight that they engaged in hoarding, ranging from good/fair appreciation to delusional perception.

Does building a museum collection qualify as hoarding? It is hard to say. Museums have many scientifically useful, aesthetically pleasing, and educational items. Our great museums have been foundational in the advancement of science, education, and in the broadening and refining the perspectives of millions of visitors, as well as nonvisitors who have benefited from their archives. Nevertheless, museums have not so much crossed the line between archiving and hoarding as obliterated it. As the saying goes, you never know when something might be useful. Why discard a perfectly serviceable shrunken head? Better to archive it. The same mindset can lead a curator to believe that it is obviously vital to preserve actual examples of present culture. One day, scientists may need to analyze a representative collection of sporks or hula hoops, for example, to understand changes in eating habits and manners, as reflected by the spork, or whether the hula hoop had any real connection to Hawaiian civilization. As the saying goes, you never know when something might be useful, even a sentence, so better to keep as many instantiations of the thing as possible, and then "curate" them. Whispered the hoarder, "That back room there? That's where I keep the <u>newspapers</u>."

Museums that nonsystematically gathered collections of thousands of human remains and put them in their back rooms and basements may or may not have been hoarding, but they had little insight into the difficulty of characterizing ancient populations based on nonsystematically collected remains. For example, an entombed Pharaoh and his retainers are unrepresentative of the ancient population of Egypt's Middle Kingdom.

A collection of ritually buried American Indian skeletons, or skeletons and skulls taken as trophies by soldiers and then donated to a museum, is again unrepresentative of the source population in ways that are not easy to evaluate. In some ways, although it would be unfair to say people with hoarding disorder are more effective than museum directors, it is better to have a back room filled with old newspapers than a museum full of skeletons, because the newspaper collection may be exhaustive, sequentially uninterrupted, consistently produced, and each page is dated, numbered, and sourced. Similarly, a telephone book issued in the year this book was written may be of greater use than this book to future historians trying to understand what was happening.

Representativeness is why certain events, such as volcanic eruptions that entomb an entire city, as at Pompeii, or undersea landslides that bury and lead to the fossilization of entire communities of organisms, are especially valuable. However, the main impulse of museum collecting of the 18th, 19th, and early 20th century appears to have been to fill museums with fine representatives of each form, attractive artifacts, and to amass as large a collection as possible. The collector's impulse is strongly expressed for objects with cultural and emotionally primal significance. Stars fallen from the sky, skulls, gems, weapons, colorful animals, monsters of the deep. At any great museum visitors, who are of the same species as curators, devote more time and attention to totemic objects than exhibits that are of equal or greater cultural, aesthetic, or scientific importance, but happen to be small or lackluster.

Several decades ago, a study of the Reptile House at the Washington National Zoo revealed that visitors behaved strangely. Diligent "zoo landers" must complete the rounds of a thousand exhibits, perhaps while accompanied by screaming children. Therefore it is not unexpected that they linger at any individual snake or lizard for an average of only about 10 s. Surprisingly, they stand before the exhibit for about 10 s, no matter the zoological, evolutionary, or ecological importance of the reptile and disregarding the care with which the exhibit was prepared. For example, who can waste precious minutes looking for a hidden animal? Move on. Nothing to see here. Except if the reptile is large or dangerous.

Instead of playing "Where's Waldo," people prefer to spend a minute studying a venomous coral snake, or better yet, a giant anaconda lying inert on a concrete slab bare except for a drainage hole. The obvious solution: procure and exhibit Komodo dragons. If one lion is impressive, why not a pride of lions, naturalistically posed on a fake kopje, the rocks on the rocky hill arranged for easy camera sight lines. Faced with a budget crunch, a Chinese zoo recently passed off a Tibetan mastiff (a dog that looks leonine) as a lion. A mother who told her child that the dog was a special type of lion lost credibility when the lion barked. Notoriously, the Houston Zoological Park (Zoo) once lost its coral snake (*Micrurus tener*, the Texas coral snake), prized because coral snakes are both colorful and deadly. It was nine months before an alert visitor noticed that the snake had been replaced with a rubber replica, which curators never even bothered to move. "We have had live snakes in the exhibit, but they don't do well – they tend to die," said a zoo spokesperson. "Rather than kill snakes, we put out a rubber one for people to be able to see what they look like" (NY Times, September 20th, 1984). Another zoo, possibly in jest, sent the Houston zoo a rubber snake, with proper breeding documentation.

At any great museum, anyone would be disappointed to view an exhibit with a rubber skull or even just one authentic human skull. Ten skulls, or better yet a pile of skulls and the reassurance that hundreds more are stored somewhere in the basement, are a different story. A pile, or array, of skulls resonates with an ancient part of our brains. The compulsion to collect is evolutionarily programmed, leading ancient peoples to bring back to their hovels and caves objects that could be made into tools, thrown, traded, or used as ritual totems. From an early age, humans, nonhuman primates, and other tool-using species (for example, tool-using crow species) are compelled to pick up and carry with them objects that may be directly used or used after modification. Many bird species, for example the bowerbird, collect and ceremonially present shiny objects to prospective mates, and several shiny objects work better than one.

Modern humans, who have so many more things they can collect and who have places to store the stuff, become obsessed with the collection of almost any object or living thing one might name: butterflies, insects, magazines, books, stamps, coins, baseball cards, gaming cards, minerals, cats, dogs, eggs, nests, watches, shoes, dresses, jackets, ties, bowties, cars, motorcycles, planes, toy soldiers, Barbie dolls, Beanie Babies, "art," small "smashable" figurines, and, of course, gold, silver, and money. Some collect words.

In conclusion, it is important to reemphasize that the scientific curation of examples of a thing, including the systematic study of modern (anthropology) and ancient (archeology) humans and of dead humans and other species (paleontology) should not be confused with mere collecting, even if at their origins these sciences were impelled or facilitated by the irrational urge to collect and display things. However, many would agree with Redman, who said," We owe it to the dead to keep better track of our prolonged efforts to turn them into trophies, scientific specimens, and valuable collectables. These significant wrongs must be addressed. We must also after so many years—finally—fully confront this complex legacy in a more thorough manner, one deserving of our most important cultural institutions." The British Museum, which "holds and cares for some 6000 human remains," has begun displaying mummies in dark cases, as a sign of respect. This is a good beginning.

http://www.britishmuseum.org/about_us/management/human_remains.aspx.

14 ◎

Birth

Before I formed thee in the belly I knew thee, and before thou camest forth out of the womb I sanctified thee.

Jeremiah 1:5.

When Henry VIII's wife felt the "quickening," celebratory bonfires were lit across England to mark the occasion of God sending England an heir to the sovereign. Were they wrong to mark that moment as the beginning of a life? This book is nearly over, but right up to this point I have artfully dodged the question of when a person begins. Skirting the timing of personhood is acceptable only to a point, because the answers are foundational to how we treat the unborn and pieces of the dead. We have taken a journey through the haploid/diploid circle of life, and the nature and potential of cells and molecules.

Science informs us that each new person begins with a cell, and that one molecule became the first cell and then all life on Earth. What science cannot disclose is when "a" human life—personhood—begins. This is a curious situation. Accredited sages of various types tell us that their lives began long before the sperm fertilized the egg from their mother, an embryo implanted in their mother's or surrogate mother's uterus, or they began kicking inside that uterus 20 weeks after gestation, which often is regarded as the moment the fetus gets a soul. Perhaps their guess is as good as anyone's. Science also tells us that human generations alternate between haploid and diploid and that all life arose from ancient precursors. Scientists do not have the answer as to when a particular human life begins.

Consistent with my main narrative that any human life is an extension and further expression of antecedents, perhaps there is no single moment of beginning. Arthur Caplan, professor of bioethics at New York University, struggled with this puzzle (*Slate*, April 4, 2017, as quoted by Elissa Strauss): "Many scientists would say they don't know when life begins. There are a series of landmark moments. The first is conception, the second is the development of the spine, the third the development of the brain, consciousness, and so on." Caplan has also written that many scientists have refrained from any comments about the beginnings of life from professional fear, and to avoid offending the powerful. For the record then, and as happens to be true, I do not feel that scientists have the answer as to when a human life begins. On the subject of when life begins, and at the risk of sounding both pontifical and obscure, science offers questions and impetus to question. But Caplan is correct to admonish scientists that it is wrong to abandon the discussion for fear of being misconstrued or being used for one agenda or another.

Immortal. https://doi.org/10.1016/B978-0-323-85692-8.00014-9

Not all scientists and physicians are reluctant to speak out. The American College of Pediatricians (March 2017) "concurs with the body of scientific evidence that corroborates that a unique human life starts when the sperm and egg bind to each other in a process of fusion of their respective membranes and a single hybrid cell called a zygote, or one-cell embryo, is created": (https://www.acpeds.org/the-college-speaks/position-statements/life-issues/when-human-life-begins). More than counterpoised against this "pro-life" view, the much better-known American Academy of Pediatrics reaffirmed an adolescent's right to confidential care when seeking abortion: https://doi.org/10.1542/peds.2016-3861 as published recently in the prestigious journal *Pediatrics*. The American Academy of Pediatrics statement is in line with other major professional medical societies, including the modern American Medical Association. These "pro-choice" medical organizations are not addressing the specific point that human life may begin at conception but are emphasizing a mother's—even an adolescent girl's—right to terminate *her* pregnancy. While some label abortion murder and others do not, no responsible medical society and probably very few pregnant women diminish the gravity of the decision to end the life of a fetus, or the context. As articulated by the Academy of Pediatrics, in the case of a pregnant adolescent wanting an abortion, the physician also has responsibility to try to facilitate family communication and to address potential realities of incest or sexual molestation that may have led to the pregnancy. However, it is fair to say that abortion is not simply "health care" with which it is sometimes equated.

Before 20 weeks, a mother can already hear her baby's heartbeat and see its recognizably mammalian face, if she wants. With ultrasound she can see into her womb, and via in vitro fertilization we can nurture human life in a Petri dish at the single-cell stage. Historian Sara Dubow in *Ourselves Unborn: A History of the Fetus in Modern America* wrote that in the late 19th century American physicians began arguing that abortion should be illegal, precisely at the time when scientists began collecting human embryos and realized that human development could be tracked back to conception, with no observable point of demarcation. In 1859, the American Medical Association (Report on Criminal Abortion, *JAMA*, Vol XII-6, 1859) unanimously approved a resolution condemning abortion at any moment following conception.

Fetal death is common, even without induced abortion, which is stunningly common. According to a fact sheet from the Guttmacher Institute https://www.guttmacher.org, some 926,200 abortions were performed in the United States in 2014, about 1.4 per 100 women aged 15–44, representing a continuing and substantial decline from the abortion high point in the 1970s, when twice as many fetuses/babies were aborted. By age 40, about 1 in 4 women will have had an abortion. Abortion is used by all ages of women and by all religions and social classes. In 2014, as reported by the Guttmacher Institute, adolescent girls accounted for 12% of abortions, Catholics for 24%, Protestants for 17%, evangelical Protestants for 13%, Whites for 39%, Blacks for 28%, Hispanics for 25%, and the poor and low income accounted for 75% of all 2014 abortions. Two-thirds of abortions were performed at 8 weeks or earlier, 88% at less than 12 weeks, and only 1.3% after 21 weeks of gestation. Increasingly, abortion is performed early and is induced by medication, mifepristone having been approved in 2000 and used in combination with misoprostol for abortion up till 10 weeks after conception and almost uniformly without risk to the physical health of the mother.

In calibrating the significance of abortion, it is relevant that it is far from certain that any fetus will be born. Death of a live-born infant is an unexpected tragedy. Fetal loss is an expected tragedy. In the United States the neonatal mortality rate is about 6 per 1000 live births. This is more than an order of magnitude lower than fetal mortality rates, fetal mortality declining with increasing fetal age. Following fertilization, and over the next several critical days, more than half of fertilized eggs fail to implant. These die. As discussed elsewhere in this book, spontaneous abortion is more the rule than the exception, and often so early that the mother is unaware she was ever pregnant. Furthermore, as discussed at more length earlier in the book, elective abortion as part of modern pregnancy planning has had the effect of very dramatically reducing the total number of abortions. These facts should not necessarily change anyone's mind, but are important as context for the abortion debate.

Increasingly—the number was 7518 in the United States in 2015—women delaying pregnancy freeze eggs for later fertilization and implantation into the uterus. Increasingly, they are freezing a lot of eggs; most clinics advise 10–20 to be frozen, because most eggs fail at one point or another along an interventional pathway that includes cryogenic freezing, fertilization, uterine implantation, and early embryonic development. Eggs from younger women do better, but the more that are frozen the greater the likelihood of a live birth—a woman freezing 10 eggs at age 36 having a 6 out of 10 chance of having a child, as of 2018 (cited by Shelly Tan, *The Washington Post*, Jan 28, 2018).

Initially the embryo undergoes a series of cell divisions until by a few days after fertilization it is a multicell ball called a morula. By day five it becomes a hollow sphere known as a blastula, at which point the three primary germ layers—ectoderm, mesoderm, and endoderm—are already present. The blastocyst invades the mother's uterine wall at about day seven, and gastrulates at about day 16 with the outer layer of the blastocyst enfolding into the central cavity, forming the blastopore. The early days of life are therefore marked by startling transitions in form, enabling ever more precise and definitive steps in differentiation toward a recognizably human fetus and baby. And all the while, the growing, dividing cells are in some ways a viable human, and in many ways, less.

Perhaps a person does not begin until gastrulation. If so, no one knows why. Perhaps a person begins later, for example 6 weeks after conception, when heartbeat can be detected. But again, no one knows why. Since 2011, heartbeat-based abortion bans have passed in several states, although they have not survived court challenges. In Roe v. Wade, the Supreme Court adopted the standard that abortion was allowable up to the point that a baby was not viable outside the womb, babies younger than 26 weeks not surviving in the decade of the 1970s. Obviously, this ruling, still in place, is a morally vacant compromise that would inevitably be embarrassed by advances in medicine. Babies as young as 22 weeks now have good odds of survival (Younge et al., 2017) and in the future it is likely that fetuses can be nurtured outside the womb entirely, and babies thereby produced.

It can be claimed that viability outside a uterus via artificial means is not the same as "independent viability" but this argument is easily deflated. What newborn infant is independently viable? None. All newborn humans are altricially helpless and dependent on their parents in a way that newborns of many other species are not, smaller and more vulnerable than the adults of their species though they may be. A sea turtle extricates itself from its eggshell, digs itself out of the sand, and immediately afterwards crawls toward the waves, facing all the perils of the beach and the great blue sea beyond, quite on its own. A newborn human is only prepared to suck on its mother's breast and cry. For abortion, the viability standard was a practical stopgap whose limitations were exposed by medical progress, but even at the beginning it was apparent that it was only a compromise of convenience.

If killing of nonindependently viable humans is acceptable, it is an unfortunate fact that a variety of killings of convenience—for example of the helpless newborn and dependent, sick and dependent, or infirm and dependent—could by extension become permissible. Independent viability is not a durable moral standard. As is easily learned by watching one or two episodes of survivor programs such as *Naked and Afraid*, the average person cannot last more than a few days in the natural environment without all kinds of tools, materials, foodstuffs, water, shelter, and psychological support provided by others. Any society deciding to rid itself of its helpless and infirm is on a path of self-annihilation. As has been proven in all ages of human history and in most or all cultures—not just Nazi Germany—when people are relieved of moral responsibility many revert, or descend, to expediency. They rid themselves of inconvenient others and construct self-justifying narratives based on false history and false science. Having rid themselves of the "worst" 10%, those who passed judgment are likely to find themselves judged at the next purge.

In Jewish religious tradition, and in some Judeo-Christian offshoots (but not all), life is said to equally begin in stages. In Exodus, it is written that the punishment for killing a person would not equally befit the killing of the fetus of a pregnant woman. However, at some point in each religious tradition, be it Judaism, Christianity, Islam, Buddhism, or other, a soul is breathed into the fetus. For Sunni Muslims, this is believed to happen at 120 days. Before 120 days, there are restrictions. After 120 days, quite logically, there are more restrictions. For any who believe in reincarnation, whether or not they grasp the consequences, the moment the soul inhabits the body should be a turning point, but surprisingly it most often is not.

From a legalistic standpoint we would like to pinpoint the moment when life begins, but from an ethical and spiritual sense it is probably better that we do not. There is a gap in knowledge and definitions, which is better to accept rather than pretend. The idea that human life begins when a baby is out of the womb and takes its first breath is patently absurd. As it is written in Jeremiah, "Before I formed thee in the belly I knew thee, and before thou camest forth out of the womb I sanctified thee." With apologies to anyone believing that human life begins after 20 weeks, at gastrulation, when the heart starts beating, or at 120 days after conception, science can offer no support, but can easily serve up contradictions to any standard. If the soul exists, science might one day detect it and measure it. Until that time, science is incapable of offering a point of demarcation for the beginning of human life except to say that everything we are came from long before, and will—in some ways large and small—go onward. In this way, every cell and molecule that preceded us and makes us possible has a special quality, deserving of attention, respect, and remembrance.

Anastasia

*I found your name in a book and in
that book a letter, were the words of
a girl who said that she was led away and
left for dead but somehow she survived it*
"I am Anastasia"—Sponge (a rock group)

Grand Duchess Anastasia Nikolaevna, at age 14yrs Born 1901, St. Petersburg, Russia Died 1918, Ekaterinburg, Russian Soviet Republic. *By Boissonnas et Eggler, St. Petersburg, Nevsky 24. Bain News Service, publisher. https://commons.wikimedia.org/w/index.php?curid¼27287143.*

Immortal. https://doi.org/10.1016/B978-0-323-85692-8.00015-0

Anna Anderson's ashes are buried in Bavaria beneath a Russian cross, and in Cyrillic, the name Anastasia and the words "Our heart is unquiet until it rests with you, Lord." After the slaughter of the Romanovs, at least 10 women claimed to be Anastasia Nikolaevna, the lone survivor of a dynasty.

Whether or not they were even Russian, and most were not, the claim of royal blood was a rich tradition of Imperial Russia. The most famous impostors to the dynastic succession were the "False Dmitris" impersonating Dmitry Ivanovich, the youngest son of Ivan the Terrible. After Ivan the Terrible's death, Boris Godunov, of opera fame, aspired to the throne, the obstacles in his way being the infertile Feodor I, who had ascended to the throne, and Dmitry, Feodor's half-brother by one of Ivan's many wives. By some twist of fate, Dmitry died of a stab wound to the throat, the finding of the official investigation being that the 8-year-old Tsarevich had the bad luck to suffer a seizure at the precise moment he was playing a game involving holding a knife to his own throat. Some historians accept this version, while others feel that Dmitry may have been assassinated by Godunov's henchmen. However, at the time, an important alternative story became popular among Polish nobles pushing impostors, but even among historians, and this narrative was that Godunov's thugs, not knowing what Dmitry looked like, murdered the wrong boy.

The most famous claimant to the identity of Anastasia was Anna Anderson, whose story was that she survived by feigning death. For more than three decades Anderson sued for official recognition in German courts, until her claim was rejected in 1970. In 1984 Anderson died and in accordance with her wishes was cremated. End of story.

Except it was not. A little technology and a quirk of genetics, the transmission of the minute genome of the mitochondrion through the maternal lineage, enabled two teams of ingenious geneticists, one led by Peter Gill and the other by television producer Maurice Philip Remy, to ask whether Anderson was really Anastasia. It was a race, Remy's group having come into the possession of a 43-year-old blood sample from Anderson, and Gill's team extracting Anderson's DNA from an intestinal biopsy sample embedded in paraffin. The race ended in a tie, with both groups of scientists the winner and Anderson the loser.

Mitochondria are small cellular organelles that are the cell's energy powerhouses. As originally proposed by Luck, mitochondria began hundreds of millions of years ago as intracellular commensal organisms capable of oxidizing organic molecules and generating ATP (adenosine triphosphate) to power other metabolic processes of cells. Mitochondria have been replicating in the cells of humans and all the eukaryotic life that preceded and coevolved with the lineage that led to humans, ever since, being transmitted from generation to generation, very much like George Lucas's mysterious Midi-chlorians. For the education of Star Trek fans, Midi-chlorians transduce the mysterious Force that binds and animates the whole universe. The Force generated by the Midi-chlorians can be used for good or evil. Well, people may not have Midi-chlorians—none have ever been identified—but mitochondria are readily detectable: some cells having high energy (force) needs are packed with thousands of them. Mitochondria divide to make more when more are needed, cause mitochondrial diseases if defective, and are closely associated with the phenomenon of programmed cell death (apoptosis), the cell reacting aversely to mitochondrial dysfunction.

Remarkably, mitochondria still have their own small, self-replicating chromosome, consisting of about 20,000 DNA nucleotides in a circular configuration. This chromosome carries a small number of genes essential to the structure and function of mitochondria, which have not, over the eons, been taken over by the cell's nuclear DNA, and this is the machinery for RNA processing and protein synthesis. Genes on the mitochondrial chromosome are transcribed within the mitochondrion, and within the mitochondrion the messenger RNAs are processed and translated into proteins using enzymes and ribosomes that are distinct from the rest of the cell. In a real sense, and not just as metaphor or evolutionary relic, mitochondria remain commensal organisms within eukaryotic cells. In most plants, but not in animals, plastids are an additional intracellular commensal, their unique function being another type of energy magic, photosynthesis.

Apart from their importance in biophysiology and medicine, mitochondria are invaluable in genetics to track familial and evolutionary relationships. Gill knew that mitochondria are transmitted maternally, from mother to offspring, with only occasional mutations over many generations. Over time, these changes to mitochondrial DNA have accumulated, producing DNA fingerprints that are characteristic of particular maternal lineages, and that can be used to determine whether two people had the same mother, grandmother, and great-grandmother. Assuming that normal genetic processes apply, and if Jesus had had a daughter, a maternal lineage descendent of Jesus would have approximately zero Jesus DNA but would have the same mitochondrial chromosome transmitted to Jesus by his mother Mary. By calibrating the rate at which mutations accumulate in the mitochondrial chromosome over the generations, it was possible to determine that all humans living on Earth today have a common maternal ancestor who lived less than 130,000 years ago. We carry genetic contributions from many other male and female ancestors, but the mitochondrial chromosome itself derives from one ancestral Eve.

The team of gene sleuths led by Peter Gill who decided to expose Anna Anderson's false claim knew that if Anastasia and other European royalty shared the same maternal lineage, they thereby should have shared the same mitochondrial DNA. One obstacle—Anna Anderson had instructed that after death her body be cremated and her wish was carried out, destroying all traces of her mitochondrial DNA. Except it did not. Biopsy tissue from Anderson was located in an Arizona hospital. Her mitochondrial DNA fingerprint was compared to that of Prince Philip, Duke of Edinburgh, the grand-nephew of Empress Alexandra, Anastasia's mother, and to DNA from the bones of Tsar Nicholas and the Tsarina, unearthed in 1991. Anderson's DNA did not match. Many people who are unrelated have the same mitochondrial DNA fingerprint, by very ancient ancestral connections. However, Gill's group found that the mitochondrial DNA pattern from Anderson's biopsy did match the DNA fingerprint of the great-nephew of one Franziska Schanzkowska, a missing Polish factory worker who as far back as 1927 had been recognized as Anderson/Anastasia by a former roommate (Rebecca Fowler, The Washington Post, 1996). Remy, whose team can be said to have shared equal credit for discovery of Anderson's true identity, summed up the situation from his perspective: "Perhaps anticipating science, Anderson requested she be cremated before her death. Since the genetic secrets of her body could not be derived from ashes, it seemed as if the mystery would never be solved, but at last we can say that this woman, who was supported by champions throughout her life, was not Anastasia."

Anna Anderson's story is thus one of pathos and posthumous disgrace, for her life and for Anastasia's and to the extent that their life stories have been written in the AGCT code of DNA, or words of books. Initially and over the years, more women came forward to claim they were Anastasia, and, even without DNA testing, it was abundantly clear most were fraudulent, deluded, or psychotic. This is similar to the puzzle of the new patient on the psychiatric ward who discovers that there are already two other people on the unit purporting to be Senator Elizabeth Warren. Anderson herself was involuntarily committed to a psychiatric hospital shortly before her death, and in a bizarre turn of events, her husband, a historian, broke her out—or abducted her—the episode ending when the Charlottesville police surrounded the escape car, rifles at the ready. Without going into the details of all claimants to the identity of Anastasia, it is clear that all could not have been Anastasia, there is no evidence that any were, and good evidence that none were.

Strangely, Professor Gill's and producer Remy's DNA tour de force exposure of Anna Anderson as a fraud was not the end of the mystery. Yes, through the technical wizardry of genetics we know that one woman was not Anastasia. However, when the burial site, or mass grave, of the Romanov family in Ekatarenburg was excavated in 1991, it was deduced that Anastasia's skeleton was missing. When the remains from the mass grave were reburied 7 years later, a skeleton corresponding to a female 5'7" tall was interred as Anastasia. However, photographs reveal Anastasia to have been shorter, and none of the skeletons recovered from the mass grave had shown the hallmarks of an immature 17-year-old, for example, undescended wisdom teeth.

The intrigue was extended in 2007, when two partially burned skeletons were found by a Russian archeologist at a bonfire site, matching an account by a witness that bodies of two of the Romanov's had been removed, burned, and buried elsewhere. The remains were of an adolescent boy and a young woman. DNA testing by independent laboratories revealed that the remains were both Romanovs. Now, and by the diligence and technological capacity of archeologists and geneticists, all 11 Romanovs, the Tsar, Tsarina, and the children, were accounted for, each with a unique but familially related DNA profile. Case closed? If so, geneticists and television producers will find other quarry.

In his last novel, *Austerlitz*, W.G. Sebald observed that just as we have appointments with the future, we have engagements with the past (James Woods, *The New Yorker*, June 5 and 12, 2017, p. 97). Humans are compelled by a mix of emotions—curiosity, nostalgia, morbidity, fear, respect—and vocation (soothsayers, geneologists, geneticists, archeologists and historians alike) to probe "what has gone before and is for the most part extinguished." The difficult-to-fulfill imperative to connect with people "on the far side of time" reminds us of George Eliot's description (in *Middlemarch*) of how harrowing it would be to be open to all the suffering of the world, "that roar which lies on the other side of silence." The voices of all those Anastasias, Terris, Nancys, Karens, Annas, Jahis, and Henriettas would surely overwhelm us if we did not somehow close our ears and minds to them. And yet, like Sebald, we are drawn to the songs of the dead, like a person who knows better than to chance fate but sometime well after the midnight hour ascends narrow stairs to investigate creaking sounds in the attic. If you have made it this far, thank you for joining me on pieces of such a journey.

In this book we have seen that single cells such as those derived from Henrietta Lacks are not miniature versions of ourselves, but their moral significance is tied both to the human they were derived from and their potential uses. Neither does Anna Anderson's DNA, or Anastasia Romanov's, or our own, reveal to us who any of these people were, or represent a continuity of being.

Epilogue

In the spring of 2020, as the nation endured COVID-19 and looked itself in the mirror to see if it upheld its most dearly held tenets of justice and equality, my family grappled with its own end-of-life/coma tragedy. Matthew Reilly, husband of my daughter Evir, suffered severe head trauma in an automobile versus pedestrian accident. As usually happens, the pedestrian suffered the worst. Matt passed away a week later in a hospital nearly locked down by the COVID-19 pandemic. So much of what I wrote about the nature of human life in this book was instantiated in the death of Matt and the days leading up to it. Matt was a high school science teacher in the prime of life. The damage to his brain echoed that suffered by Quinlan, Cruzan, Schiavo, and McMath, and had the potential to raise the same moral dilemmas if he had transitioned from coma to persistent vegetative state. Several days before Matt died, and fortunately only for a horrible few hours, my daughter and through her each of us was given false hope by a well-meaning ICU nurse who had misinterpreted pathological signs, as is all too easy to do after a brain has been severely damaged. Doctors joined together with our family, namely Evir, Matt's parents Cindy and John, and my wife, Nadia, who is a neurologist, to walk the family step-by-step through the barren, end-of-life realities of Matt's injuries, and prognosis. Matt's doctors, and Matt's loving family rallying around each other, helped the family make the best decision that could be made on Matt's behalf. Physical parts of Matt—his heart, liver, and kidneys—now sustain other lives. His unique personality, intellect, and humor live on in us.

Notes

Tristram Engelhardt: Author of Foundations of Bioethics and a philosopher and bioethicist wrote that "What distinguishes persons is their capacity to be self-conscious, rational, and concerned with worthiness of blame and praise. The possibility of such entities grounds the possibility of the moral community. It offers us a way of reflecting on the rightness and wrongness of actions and the worthiness or unworthiness of actors."

Balancing the books, Engelhardt wrote:

On the other hand, not all humans are persons. Not all humans are self-conscious, rational, and able to conceive of the possibility of blaming and praising. Fetuses, infants, the profoundly mentally retarded, and the hopelessly comatose provide examples of nonpersons. Such entities are members of the human species. They do not in and of themselves have standing in the moral community. They cannot blame or praise or be worthy of blame or praise. They are not prime participants in the moral endeavor. Only persons have that status.

Embracing Engelhardt's formulation of personhood, consider how we treat the profoundly mentally deficient or the hopelessly comatose—with respect and dignity. Not as if they were deserving of blame or praise, but as morally significant. We need not be concerned about the autonomy or dignity of the cells, DNA, or individual genes except as our actions impinge on our respect for persons. They do. We have to consider their derivation and connection to the still-living memory. The distinction between a human body, with its billions of cells, and a human or society formed of humans, is brought home by considerations of individual autonomy and dignity of unit constituents, and the ability of a person to consent.

It would be silly, condescending, or an act of false compassion to ask the permission of entities such as cells and selfish genes that are unequipped to feel pain or desire and unable to choose, except so far as their perceptions are decoded and their choices are made by a brain. However, it is right and necessary to obtain the consent of a person for the use of any part of themselves. If circumstances require sacrifice that cannot be refused, we should at least, as in Stravinsky's barbaric rite of spring or the urbane workings of committees that clean up ethical messes, invest drama and ceremony into the process by which the person who was unfairly used gave their life or some part of themselves for the good of the whole. Science has not quite given us dominion over death and life, and probably never will, in the sense desired by those who grasp at immortality, an eternity in heaven, infinite reincarnation, or a mystical union with the universe or some other consciousness. As far as is known, séances do not work, and the dead do not return as ghosts, benign or malevolent, but there are ways more mundane and actually effective for those who follow the dead to connect to them. When we remember the dead with compassion, their lives, otherwise so fragile and brief, partly elude oblivion.

References

Abbot, P., Abe, J., Alcock, J., et al., 2011. Inclusive fitness theory and eusociality. Nature 471, E1–E4.

Adam, P.A.J., Ratha, N., Rohiala, E., et al., 1973. Cerebral oxidation of glucose and D-beta hydroxy, butyrate in the isolated perfused human head. Trans. Am. Pediatr. Soc. 309, 81.

Aviv, R., 2018. The death debate. New Yorker, 30–41. Feb 5.

Bass, J., Lazar, M.A., 2016. Circadian time signatures of fitness and disease. Science 6315, 994–999.

Beecher, H., et al., 1968. A definition of irreversible coma: report of the Ad Hoc Committee of the Harvard Medical School to Examine the Definition of Brain Death. JAMA 205, 337–340.

Berson, D.M., 2003. Strange vision: ganglion cells as circadian photoreceptors. Trends Neurosci. 26, 314–320.

Bianconi, E., Piovesan, A., Facchin, F., Beraudi, A., Casadei, R., Frabetti, F., Vitale, L., Pelleri, M.C., Tassani, S., Piva, F., Perez-Amodio, S., Strippoli, P., Canaider, S., 2013. An estimation of the number of cells in the human body. Ann. Hum. Biol. 40 (6), 463–471.

Bouchard T.J., Jr., McGue, M., 1981. Familial studies of intelligence: a review. Science 212, 1055–1059.

Brady, S.G., 2003. Evolution of the army ant syndrome: the origin and long-term evolutionary stasis of a complex of behavioral and reproductive adaptations. Proc. Natl. Acad. Sci. U. S. A. 100 (11), 6575–6579.

Brady, W.J., Gurka, K.K., Mehring, B., Peberdy, M.A., O'Connor, R.E., 2011. In-hospital cardiac arrest: Impact of monitoring and witnessed event on patient survival and neurologic status at hospital discharge. Resuscitation 82 (7), 845–852.

Carlson, B.M., 2004. Human Embryology and Developmental Biology, third ed. Mosby-Elsevier, Philadelphia, pp. 2, 8–10, 31.

Cattell, R.B., Nesselroade, J.R., 1967. Likeness and completeness theories examined by Sixteen Personality Factor measures on stably and unstably married couples. J. Pers. Soc. Psychol. 7 (4, Pt.1), 351–361.

Chagnon, N.A., 1968. Yanomamö: The Fierce People. Holt, Rinehart & Winston, New York.

Collinge, J., Whitfield, J., McKintosh, E., Beck, J., Mead, S., Thomas, D.J., Alpers, M.P., 2006. Kuru in the 21st century—an acquired human prion disease with very long incubation periods. Lancet 367 (9528), 2068–2074.

Daly, M., Wilson, M., 1996. Violence against stepchildren. Curr. Dir. Psychol. Sci. 5, 77–81.

Damasio, A.P., 1994. Descartes' Error. G.P. Putnam.

Dean, M., Carrington, M., Winkler, C., Huttley, G.A., Smith, M.W., Allikmets, R., et al., 1996. Genetic restriction of HIV-1 infection and progression to AIDS by a deletion allele of the CKR5 structural gene. Hemophilia Growth and Development Study, Multicenter AIDS Cohort Study, Multicenter Hemophilia Cohort Study, San Francisco City Cohort, ALIVE Study. Science 273 (5283), 1856–1862. https://doi.org/10.1126/science.273.5283.1856.

Diaz-Espinoza, R., Morales, R., Concha-Marambio, L., Moreno-Gonzalez, I., Moda, F., Soto, C., 2018. Treatment with a non-toxic, self-replicating anti-prion delays or prevents prion disease in vivo. Mol. Psychiatry 23 (3), 777–788. https://doi.org/10.1038/mp.2017.84.

Dyble, M., Salali, G.D., Chaudhary, N., Page, A., Smith, D., Thompson, J., Vinicius, L., Mace, R., Migliano, A.B., 2015. Human behavior. Sex equality can explain the unique social structure of hunter-gatherer bands. Science 348 (6236), 796–798.

Eccles, J., 1953. The Neurophysiological Basis of Mind. Oxford University Press, Oxford.

Einstein, A., Podolsky, B., Rosen, N., 1935. Can quantum-mechanical description of physical reality be considered complete? Phys. Rev. 47, 777.

Ferris, J.P., Hill Jr., A.R., Liu, R., Orgel, L., 1996. Synthesis of long prebiotic oligomers on mineral surfaces. Nature 381, 59–61.

Gold, M., 1986. A Conspiracy of Cells: One Woman's Immortal Legacy and the Medical Scandal It Caused. Albany University State Press, ISBN: 978-0-88706-099-1.

Goldman, D., 2012. Our Genes, Our Choices. Elsevier.

Guerrier-Takada, C., Gardiner, K., Marsh, T., Pace, N., Altman, S., 1983. The RNA moiety of ribonuclease P is the catalytic subunit of the enzyme. Cell 35, 649–857.

Haverty, M.I., Thorne, B.L., 1989. Agonistic behavior correlated with hydrocarbon phenotypes in damp-wood termites, Zootermopsis (Isoptera: Termopsidae). J. Insect. Behav. 2, 523–543.

Hesketh, T., Zhu, W.X., 1997. Health in China. Traditional Chinese medicine: one country, two systems. BMJ 315 (7100), 115–117.

Hinegardner, R.T., Engelberg, J., 1963. Rationale for a universal genetic code. Science 142, 1083–1085.

Ho, J.M.L., Bennett, M.R., 2018. Improved memory devices for synthetic cells. Science 360, 150–153.

Hou, C., Kaspari, M., Vander Zanden, H.B., Gillooly, J.F., 2010. Energetic basis of colonial living in social insects. Proc. Natl. Acad. Sci. U. S. A. 107 (8), 3634–3638.

Hughes, W.O., Oldroyd, B.P., Beekman, M., Ratnieks, F.L., 2008. Ancestral monogamy shows kin selection is key to the evolution of eusociality. Science 320 (5880), 1213–1216.

Huxley, T.H., 1874. On the hypothesis that animals are automata, and its history. Fortnightly Rev. 16, 555–580. Reprinted in Huxley, T.H., 1898. Method and Results: Essays. D. Appleton and Company, New York.

Jennett, B., 2002. The Vegetative State: Medical Facts, Ethical and Legal Dilemmas. Cambridge University Press.

Jennett, B., Plum, F., 1972. Persistent vegetative state after brain damage. A syndrome in search of a name. Lancet 1, 734–737.

Johnston, W.K., Unrau, P.J., Lawrence, M.S., Glasner, M.E., Bartel, D.P., 2001. RNA-catalyzed RNA polymerization: accurate and general RNA-templated primer extension. Science 292 (5520), 1319–1325.

Joyce, G.F., Orgel, L.E., 2006. In: Gesteland, R.F., Cech, T.R., Atkins, J.F. (Eds.), The RNA World. Cold Spring Harbor Laboratory Press, pp. 23–56.

Karvonen, M.K., Pesonen, U., Koulu, M., Niskanen, L., Laakso, M., Rissanen, A., Dekker, J.M., Hart, L.M., Valve, R., Uusitupa, M.I., 1998. Association of a leucine(7)-to-proline(7) polymorphism in the signal peptide of neuropeptide Y with high serum cholesterol and LDL cholesterol levels. Nat. Med. 4 (12), 1434–1437.

Krupovic, M., Koonin, E.V., 2017. Multiple origins of viral capsid proteins from cellular ancestors. PNAS 114 (12), E2401–E2410.

Kuo, L.E., Kitlinska, J.B., Tilan, J.U., Li, L., Baker, S.B., Johnson, M.D., Lee, E.W., Burnett, M.S., Fricke, S.T., Kvetnansky, R., Herzog, H., Zukowska, Z., 2007. Neuropeptide Y acts directly in the periphery on fat tissue and mediates stress-induced obesity and metabolic syndrome. Nat. Med. 13 (7), 803–811. Erratum in: Nat. Med. 2007 13(9), 1120.

Laureys, S., Faymonville, M.E., Peigneux, P., et al., 2002. Cortical processing of noxious somatosensory stimuli in the persistent vegetative state. NeuroImage 17 (2), 732–741.

Ledoux, J., 2003. Synaptic Self: How Our Brains Become Who We Are. Penguin.

Lincoln, T.A., Joyce, G.F., 2009. Self-sustained replication of an RNA enzyme. Science 323 (5918), 1229–1232.

Lyons, T., Reinhard, C., Planavsky, N., 2014. The rise of oxygen in Earth's early ocean and atmosphere. Nature 506, 307–315.

Machado, C., Shewmon, D.A., 2004. Brain Death and Disorders of Consciousness. Springer Science & Business Media. 288 p.

Mai, C.T., Kucik, J.E., Isenburg, J., Feldkamp, M.L., Marengo, L.K., Bugenske, E.M., Thorpe, P.G., Jackson, J. M., Correa, A., Rickard, R., Alverson, C.J., Kirby, R.S., for the National Birth Defects Prevention Network, 2013. Selected birth defects data from population-based birth defects surveillance programs in the United States, 2006 to 2010: Featuring trisomy conditions. Birth Defects Res. A: Clin. Mol. Teratol. 97, 709–725.

Massung, R.F., Liu, L.I., Qi, J., Knight, J.C., Yuran, T.E., Kerlavage, A.R., Parsons, J.M., Venter, J.C., Esposito, J. J., 1994. Analysis of the complete genome of smallpox variola major virus strain Bangladesh-1975. Virology 201 (2), 215–240.

McEwen, B.S., 1998. Protective and damaging effects of stress mediators. Seminars in Medicine of the Beth Israel Deaconess Medical Center. N. Engl. J. Med. 338, 171–179.

Meaney, M.J., Aitken, D.H., Bodnoff, S.R., Iny, L.J., Tatarewicz, J.E., Sapolsky, R.M., 1985. Early postnatal handling alters glucocorticoid receptor concentrations in selected brain regions. Behav. Neurosci. 99 (4), 765–770.

Metzinger, T., 2003. Being No One. Basic Books.

Metzinger, T., 2009. The Ego Tunnel: The Science of the Mind and the Myth of the Self. Basic Books.

Moore, K.L., Persaud, T.V.N., 2003. The Developing Human, seventh ed. Saunders-Elsevier, Philadelphia, p. 31.

Moorjani, P., Sankararaman, S., Fu, Q., Przeworski, M., Patterson, N., Reich, D., 2016. A genetic method for dating ancient genomes provides a direct estimate of human generation interval in the last 45,000 years. Proc. Natl. Acad. Sci. 113, 5652–5657.

Moseley, C. (Ed.), 2010. Atlas of the World's Languages in Danger, third ed. UNESCO Publishing. Online version, Paris. http://www.unesco.org/culture/en/endangeredlanguages/atlas.

Nirenberg, M.W., Jones, W., Leder, P., Clark, B.F.C., Sly, W.S., Pestka, S., 1963. On the coding of genetic information. Cold Spring Harb. Symp. Quant. Biol. 28, 549–557 (Google Scholar).

Nowak, M., Highfield, R., 2011. SuperCooperators: Altruism, Evolution, and Why We Need Each Other to Succeed. Free Press.

Pagel, M., Atkinson, Q.D., Calude, A.S., Meade, A., 2013. Ultraconserved words point to deep language ancestry across Eurasia. Proc. Natl. Acad. Sci. 110, 8471–8476.

Parkhill, J., Wren, B.W., Thomson, N.R., Titball, R.W., Holden, H.T., Prentice, M.B., Sebaihia, M., James, K.D., Churcher, C., Mungall, K.L., Baker, S., Basham, D., Bentley, S.D., Brooks, K., Cerdeño-Tárraga, A.M., Chillingworth, T., Cronin, A., Davies, R.M., Davis, P., Dougan, G., Feltwell, T., Hamlin, N., Holroyd, S., Jagels, K., Karlyshev, A.V., Leather, S., Moule, S., Oyston, P.C., Quail, M., Rutherford, K., Simmonds, M., Skelton, J., Stevens, K., Whitehead, S., Barrell, B.G., 2001. Genome sequence of Yersinia pestis, the causative agent of plague. Nature 413 (6855), 523–527.

Petri, R.J., 1887. Eine kleine Modifikation des Koch'schen Plattenverfahrens (A small modification of Koch's plate method). Centralblatt für Bakteriologie und Parasitenkunde 1, 279–280.

Pollan, M., 2018. How to Change Your Mind. Penguin Press.

Powner, M.W., Gerland, B., Sutherland, J.D., 2009. Synthesis of activated pyrimidine ribonucleotides in prebiotically plausible conditions. Nature 459, 239–242.

Quinlan, J., 2005. My Joy, My Sorrow: Karen Ann's Mother Remembers. St. Anthony Messenger Press, Cincinnati, Ohio, ISBN: 978-0-86716-663-7.

Quinlan, J., Quinlan, J., Battelle, P., 1977. Karen Ann: the Quinlans Tell Their Story. Doubleday, Garden City, New York, ISBN: 978-0-385-12666-3.

Ramachandran, V., 2012. The Tell-Tale Brain: A Neuroscientist's Quest for What Makes Us Human. Norton, New York.

Rebecca, J., 1996. Anastasia: the mystery resolved. Wash. Post. October 6.

Redman, S.J., 2016. Bone Rooms: From Scientific Racism to Human Prehistory in Museums. Harvard University Press.

Rettenmeyer, C.W., Rettenmeyer, M.E., Joseph, J., et al., 2011. The largest animal association centered on one species: the army ant *Eciton burchellii* and its more than 300 associates. Insect. Soc. 58, 281–292.

Rice, W.R., 2018. The high abortion cost of human reproduction. Biorxiv. https://www.biorxiv.org/content/10.1101/372193v1. https://doi.org/10.1101/372193.

Rovera, G., O'Brien, T.G., Diamond, L., 1979. Induction of differentiation in human promyelocytic leukemia cells by tumor promoters. Science 204, 868–870.

Rousset, F., Lion, S., 2011. Much ado about nothing: Nowak et al.'s charge against inclusive fitness theory. J. Evol. Biol. 24 (6), 1386–1392.

Schiff, N.D., Ribary, U., Moreno, D.R., et al., 2002. Residual cerebral activity and behavioural fragments can remain in the persistently vegetative brain. Brain 125, 1210–1234.

Schiff, N.D., Giacino, J.T., Kalmar, K., Victor, J.D., Baker, K., Gerber, M., Fritz, B., Eisenberg, B., Biondi, T., O'Connor, J., Kobylarz, E.J., Farris, S., Machado, A., McCagg, C., Plum, F., Fins, J.J., Rezai, A.R., 2007. Behavioural improvements with thalamic stimulation after severe traumatic brain injury. Nature 448 (7153), 600–603.

Shay, J.W., Wright, W.E., 2000. Hayflick, his limit, and cellular ageing. Nat. Rev. Mol. Cell Biol. 1 (1), 72–76.

Shewmon, D.A., 1997. Recovery from "brain death": a neurologist's apologia. Linacre Q. 64 (1), 30–96. Available at: http://epublications.marquette.edu/lnq/vol64/iss1/4.

Silk, J., 1980. Adoption and kinship in Oceania. Am. Anthropol. 82, 799–820.

Skloot, R., 2010. The Immortal Life of Henrietta Lacks. Crown/Random House, New York, ISBN: 978-1-4000-5217-2.

Snyder-Mackler, N., Sanz, J., Kohn, J.N., Brinkworth, J.F., Morrow, S., Shaver, A.O., Grenier, J.C., Pique-Regi, R., Johnson, Z.P., Wilson, M.E., Barreiro, L.B., Tung, J., 2016. Social status alters immune regulation and response to infection in macaques. Science 354 (6315), 1041–1045.

Stephens, J.C., Reich, D.E., Goldstein, D.B., Shin, H.D., Smith, M.W., Carrington, M., Winkler, C., Huttley, G. A., Allikmets, R., Schriml, L., Gerrard, B., Malasky, M., Ramos, M.D., Morlot, S., Tzetis, M., Oddoux, C., di Giovine, F.S., Nasioulas, G., Chandler, D., Aseev, M., Hanson, M., Kalaydjieva, L., Glavac, D., Gasparini, P., Kanavakis, E., Claustres, M., Kambouris, M., Ostrer, H., Duff, G., Baranov, V., Sibul, H., Metspalu, A., Goldman, D., Martin, N., Duffy, D., Schmidtke, J., Estivill, X., O'Brien, S.J., Dean, M., 1998 Jun. Dating the origin of the CCR5-Delta32 AIDS-resistance allele by the coalescence of haplotypes. Am. J. Hum. Genet. 62 (6), 1507–1515. https://doi.org/10.1086/301867.

Sterling, P., Eyer, J., 1988. Allostasis: a new paradigm to explain arousal pathology. In: Fisher, S., Reason, J.T. (Eds.), Handbook of Life Stress, Cognition, and Health. Wiley, Chichester, NY, ISBN: 9780471912699.

Sun, N., Youle, R.J., Finkel, T., 2016. The mitochondrial basis of aging. Mol. Cell 61 (5), 654–666. https://doi.org/10.1016/j.molcel.2016.01.028.

Szell, A.Z., Bierbaum, R.C., Hazelrigg, W.B., Chetkowski, R.J., 2013. Live births from frozen human semen stored for 40 years. J. Assist. Reprod. Genet. 30 (6), 743–744.

Tooley, M., 1972. Abortion and infanticide. Philos. Public Aff. 2, 37–65.

Van Etten, J.L., 2011. Giant viruses. Am. Sci. 99 (4), 304–311. https://doi.org/10.1511/2011.91.304. Archived from the original on 2011-06-11.

Williams, G.C., 1957. Pleiotropy, natural selection, and the evolution of senescence. Evolution 11, 398–411.

Wochner, A., James, A., Coulson, A., Holliger, P., 2011. Ribozyme-catalyzed transcription of an active ribozyme. Science 332 (6026), 209–212.

Woese, C.R., Hinegardner, R.T., Engelberg, J., 1964. Universality in the genetic code. Science 144, 1030–1031.

Wood, J.W., 1994. Dynamics of Human Reproduction: Biology, Biometry, Demography. Transaction Publishers.

Younge, N., Goldstein, R.F., Bann, C.M., Hintz, S.R., Patel, R.M., Smith, P.B., Bell, E.F., Rysavy, M.A., Duncan, A.F., Vohr, B.R., Das, A., Goldberg, R.N., Higgins, R.D., Cotton, C.M., Eunice Kennedy Shriver National Institute of Child Health and Human Development Neonatal Research Network, 2017. Survival and neurodevelopmental outcomes among periviable infants. N. Engl. J. Med. 376 (7), 617–628.

Zhang, H., Peccoud, J., Xu, M., et al., 2020. Horizontal transfer and evolution of transposable elements in vertebrates. Nat. Commun. 11, 1362. https://doi.org/10.1038/s41467-020-15149-4.

Zhelezinskaia, I., Kaufman, A., Farquhar, J., Cliff, J., 2014. Large sulfur isotope fractionations associated with Neoarchean microbial sulfate reduction. Science 346, 742–744.

Zhou, Z., Zhu, G., Hariri, A.R., Enoch, M.A., Scott, D., Sinha, R., Virkkunen, M., Mash, D.C., Lipsky, R.H., Hu, X.Z., Hodgkinson, C.A., Xu, K., Buzas, B., Yuan, Q., Shen, P.H., Ferrell, R.E., Manuck, S.B., Brown, S.M., Hauger, R.L., Stohler, C.S., Zubieta, J.K., Goldman, D., 2008. Genetic variation in human NPY expression affects stress response and emotion. Nature 452 (7190), 997–1001.

Index

Note: Page numbers followed by *f* indicate figures.

Printed in the United States
by Baker & Taylor Publisher Services